FORTY YEARS OF
CALEDONIAN LOCOMOTIVES 1882-1922

Other books on Scottish locomotives

John Thomas *North British Atlantics*
John Thomas *The Springburn Story*
Thomas Middlemass *Mainly Scottish Steam*
O. S. Nock *The Caledonian Dunalastairs*

By the same author
William Stroudley—Craftsman of Steam

FORTY YEARS OF CALEDONIAN LOCOMOTIVES 1882–1922

H. J. C. CORNWELL

DAVID & CHARLES
NEWTON ABBOT   LONDON
NORTH POMFRET (VT)   VANCOUVER

ISBN 0 7153 65339

© H. J. C. Cornwell, 1974

All rights reserved. No part of this publication may be reproduced, stored, in a retrieval system or transmitted in any form or by any means electronic, mechanical, photocopying, recording or otherwise without the prior permission of David & Charles (Holdings) Limited

Set in Linotype 8 on 9 pt and 10 on 12 pt Times Roman and printed in Great Britain at the St Ann's Press, Park Road, Altrincham, Cheshire, WA14 5QQ
for David & Charles (Holdings) Limited, South Devon House, Newton Abbot, Devon

Published in the United States of America by David & Charles Inc., North Pomfret, Vermont 05053 USA

Published in Canada by Douglas, David & Charles Limited, 3645 McKechnie Drive, West Vancouver BC

CONTENTS

		page
	LIST OF ILLUSTRATIONS	7
	INTRODUCTION	11
1	THE CALEDONIAN IN 1882 The True Line - The locomotive problem	13
2	BASIC DRUMMOND DESIGNS The 294 class (Jumbos) - The 66 class - Early tank engines - Perspective	18
3	IMPROVING THE BREED Exhibition engines - The 'coast bogies' Tenders - Tank engines 1887-90 - High pressure trials	29
4	THE TRANSITIONAL PHASE 1891-5 Personnel changes - 4–4–0 design and performance - 0–6–0 design and construction - Tank engines - Renewals	49
5	THE RISE OF J. McINTOSH The new star - The last Jumbos - Tank engines 1895-1900 - Dunalastair	68
6	THE McINTOSH STANDARD CLASSES 1897-1900 The new era - The 766 class - The 812 class - The 900 Class - Standard tank engines	78
7	McINTOSH DEVELOPMENTS 1901-4 The 600 class - New boilers - The 55 class - The 49 class - The 492 class - The 140 class - The keystone of the arch	95

8	CLIMAX AND DECLINE	117
	Cardean - The 918 and 908 classes - Quiescence	
9	THE COMING OF SUPERHEATING	136
	Reflation - The 139, 117 and 40 classes - Superheated goods engines - Tank engines 1910-14 - Assessment	
10	WILLIAM PICKERSGILL	155
	Metamorphosis - The River class - The 4–4–0s - The 60 class	
11	WAR AND PEACE	171
	New goods engines - Tank engines 1915-22 - The latter-day scene - The last generation	
	APPENDICES	187
	BIBLIOGRAPHY	210
	INDEX	215

LIST OF ILLUSTRATIONS

PLATES

	page
Drummond 0–6–0 No 306 (Photomatic	33
Drummond 0–4–4T No 1223 at Perth (Collection of A. G. Ellis	33
Drummond 4–4–0 No 90 with Drummond 3130-gallon tender (Collection of A. G. Ellis	34
Dunalastair No 723 (Photomatic	34
Lambie 0–6–0 No 335 with Lambie 2840-gallon tender (Photomatic	51
766 class 4–4–0 No 769 (Photomatic	51
Lambie 4–4–0T No 2 in original blue livery (Author's collection	52
19 class 0–4–4T No 19 as repainted black (Alastair Harper	52
'Coast Bogie' No 80 with 2840-gallon tender of late Drummond design (Real Photographs	69
Drummond 4–4–0 No 70 as rebuilt with large boiler (Real Photographs	69
439 class 0–4–4T No 222 (LCGB Ken Nunn Collection	70
782 class 0–6–0T No 245 (Photomatic	70
600 class 0–8–0 No 604 (Collection of A. G. Ellis	87
652 class 0–6–0 No 423 (Author's collection	87
55 class 4–6–0 No 59 (Real photographs	88
918 class 4–6–0 No 919 at Carlisle (Real Photographs	88
49 class 4–6–0 No 50 *Sir James Thompson* (Collection of A. G. Ellis	105

Cardean outside Crewe North shed, June 1909 (British Railways	105
908 class 4–6–0 No 911 *Barochan* (LCGB Ken Nunn collection	106
Cardean outside Polmadie shed (I. M. Coonie	106
908 class 4–6–0 No 917 (Photomatic	123
179 class 4–6–0 No 184 (Collection of A. G. Ellis	123
60 class 4–6–0 No 63 (Real Photographs	124
300 class 0–6–0 No 313 (Real Photographs	124
60 class 4–6–0 No 62 (Author's collection	141
4–6–2T No 944 (Collection of A. G. Ellis	141
956 class 4–6–0 No 959 (Collection of A. G. Ellis	142
191 class 4–6–0 No 192 (Collection of A. G. Ellis	142
Lambie 4–4–0 No 18 (LCGB Ken Nunn collection	159
956 class 4–6–0 No 957 (Collection of A. G. Ellis	159
Drummond 4–4–0 No 66 as rebuilt with large boiler (Real photographs	160
River class 4–6–0 No 941 (Real photographs	160

LINE ILLUSTRATIONS IN THE TEXT

		page
1	Arrangement of standard 18in x 26in cylinders: 1883	19
2	Footplate layout of 171 class	25
3	Footplate layout of No 124	31
4	Cylinder arrangement: batch Y21	44
5	Arrangement of standard 18in x 26in cylinders: 1891	53
6	Blastpipe variation: *Left* batch Y29 *Right* batch Y30	58
7	Connection between Westinghouse and vacuum driver's brake valves	66
8	Arrangement of standard 19in x 26in cylinders	79
9	General arrangement of McIntosh standard 0–6–0	84
10	General arrangement of standard 3000-gallon tender	85
11	Arrangement of firebox faceplate: *Left* 49 class *Right* 903 class	102
12	Smokebox arrangement of 49 class as originally built	103
13	Smokebox arrangement of 49 class as altered	109
14	903 class: section through crank-axle showing dished wheels	118
15	903 class: section through firebox showing direct crown stays	120
16	104 class blastpipe: 1904	131

17	140 class: draughting arrangement as originally designed	133
18	140 class: spark arrestor and altered draughting arrangement	134
19	Footplate layout of 117 class	139
20	60 class end elevation	168
21	Pickersgill balanced slide valve	173
22	956 class end elevation	180
23	956 class: arrangement of derived valve gear (John L. Anderson)	181
24	Footplate layout of 956 class	182

The drawings of the Drummond cylinders are the work of the late C. M. Keiller and are reproduced here by the kind permission of the Editor of the *SLS Journal*. To all of the above, the author wishes to express his thanks and indebtedness.

INTRODUCTION

When, at the age of nine, I was given a copy of A. B. MacLeod's *The McIntosh Locomotives of the Caledonian Railway* I had already developed a keen interest in the Caledonian engines, then drab in LMS wartime livery. But MacLeod's book portrayed the locomotives in what for me was a new and romantic light and I soon found myself searching for more information. Invaluable at this period was J. F. McEwan's "The Locomotives of the Caledonian Railway" published in *The Locomotive* between 1940 and 1948. Later on I was able to consult *The Engineer, Engineering* and *The Railway Engineer*, each of which contained detailed descriptions and drawings of various classes. The writings of A. G. Dunbar in *Trains Illustrated, Railway World* and the *SLS Journal* provided much information on the allocation and workings of individual engines and painted a picture of the various classes from the viewpoint of the men who handled them. The articles written by A. J. S. Paterson for the same journals also added much flavour to the subject. As time passed, I got to know these authors well and it is with pleasure that I acknowledge their help. Mr Montague Smith also provided much assistance and I acknowledge my indebtedness to him. Other enthusiasts who provided information were Mr David L. Smith, Dr Neil MacKillop, Mr Albert Grieg and Mr Graham King.

My personal research began through the good offices of Mr J. Snodgrass, then chief draughtsman for the Scottish Region of British Railways, when I spent a fascinating afternoon going through the drawing office at St Rollox under the guidance of Mr William Adams, who had begun work there just after World War I. An important consequence of this visit was that I obtained the

INTRODUCTION

numbers of the general arrangement drawings of most Caledonian classes. The next step was to purchase a print of each of these drawings with the help of Mr Adams (Drawing Office) and Mr W. Sharp (Public Relations Office). When the time came for the drawing office to dispose of its steam locomotive drawings, a large number of Scottish subjects were presented to the Glasgow Museum of Transport. Through the generosity of Mr A. S. E. Browning, Curator of Technology, and Mr James L. Wood, his assistant, and with the permission of Mr G. S. Calder, the Chief Mechanical and Electrical Engineer of British Rail's Scottish Region, I was allowed to acquire a number of drawings surplus to the requirements of the museum. This enabled me to build up a very large dossier on Caledonian locomotive design.

But to have concentrated on design alone would have produced a one-sided view of Caledonian locomotive history. To supplement the technical detail which was now at my disposal, I turned to the company minute books, working timetables and locomotive department records in the Scottish Record Office, Edinburgh. I received much help from Assistant Keeper and fellow railway enthusiast George Barbour. Don Martin and Murdoch Nicholson of the Mitchell Library, Glasgow, gave similar help by looking out documents and photographs from the records of the North British Locomotive Company. My sincere thanks to them all.

I must also thank Alan G. Dunbar and Montague Smith for their help in checking my manuscript. Each went to considerable pains over it and I am grateful to them both. Last but not least, I must acknowledge my debt to John Thomas, without whose encouragement and advice this book would never have been written.

CHAPTER 1

THE CALEDONIAN IN 1882

THE TRUE LINE

The Caledonian Railway was a national institution in which the Scottish people took a pride. With its blue locomotives, each carrying the royal arms of Scotland, its distinctive chocolate and white carriages, its elegant stations and its yacht-like steamers, it well deserved its self-assumed title, 'The True Line'. It was not always so; before 1883 it had had a chequered history. But over the succeeding 20 years it was transformed by the leadership, drive and ability of some dozen men. One of these was Dugald Drummond.

By 1882 the company's locomotive stock had fallen short of requirements and its workshops at St Rollox, essentially unchanged since the days of Robert Sinclair, were unable to cope with further increases in stock. George Brittain, the locomotive superintendent, was a 'running' man rather than a designer. Moreover he was a sick man. It was a delicate situation. Correspondence between the general manager and the locomotive superintendent was read to the board on 21 March 1882 and instructions were given to the locomotive and stores committee to enquire into the management of the locomotive department and take such action as it thought expedient. What the committee thought was outlined in a board minute of 25 April:—

> The Committee reported that in consequence of the readjustment of Mr Brittain's department from the increase in work, they had to recommend the appointment of a Locomotive Superintendent. Mr Brittain's salary in the meantime to remain as at present and his duties for the future to be hereafter adjusted.

THE CALEDONIAN IN 1882

Nominally, Brittain was being retained as a consultant but in reality he was being demoted.

The Caledonian board was aware of the transformation that had taken place in the North British locomotive department following the appointment of Dugald Drummond as locomotive superintendent in 1875. By 1882, Drummond had acquired the highest reputation as a locomotive designer and it is possible that the Caledonian directors set out to capture his allegiance. According to legend, Drummond found himself seated at the same hotel table as two Caledonian directors and accepted the post of locomotive superintendent before the first course was finished. Negotiations must have been going on surreptitiously, for the appointment was not mentioned in the board minutes until 20 June, by which time it was a *fait accompli*. 'Reported engagement of Mr D. Drummond as Locomotive Superintendent of the Company at a Salary at the rate of £1700 a year, to date from the day on which he enters the duties of the office'. The salary was double that of Brittain's—evidence of the board's determination to get their man.

Drummond began work on 14 August and met the locomotive and stores committee for the first time eight days later. On 27 November, Brittain informed the board of his wish to retire and was superannuated from 9 December.

All important matters, particularly those involving capital expenditure, had to have the approval of the full board of directors, but control over the working of the departments was exercised by the committees, the affairs of the locomotive department being dealt with by the locomotive and stores, and the traffic and permanent way committees. The first was responsible for the purchase of materials and components for locomotive construction and repair, while the second was concerned with the provision and operation of locomotives as well as with improvements in shed facilities. On technical matters, the locomotive superintendent was responsible to the locomotive and stores committee, whose weekly or fortnightly meetings he attended. Unlike the general manager who attended the board meetings *ex officio*, the locomotive superintendent met the full board only when specifically invited to do so. As the company's chief officer the general manager acted as intermediary between the board and the heads of the departments, the locomotive superintendent being wholly subservient to him except in those technical matters for which the locomotive and stores committee accepted responsibility.

The locomotive department was divided into an outdoor department governing the allocation and running of locomotives headed by the assistant (outdoor) locomotive superintendent, John Lambie, and an indoor department concerned with the design, construction and repair of locomotives and rolling stock, presided over by Joseph Goodfellow. With Drummond's arrival the staff of the indoor side was reinforced by a number of experienced Cowlairs men, including George Thomas Wheatley, William Montgomerie Urie and Peter Drummond. Another outsider, Robert Wallace Urie, became chief draughtsman about 1885, this office being second only to that of works manager.

THE LOCOMOTIVE PROBLEM

The 257 goods engines built during the preceding 20 years were of four basic types:—

(1) 6ft 2in 2–4–0s   107 engines in four classes
(2) 5ft 2in 2–4–0s   17 engines in one class
(3) 5ft 2in 0–4–2s   94 engines in two classes
(4) 5ft 2in 0–6–0s   39 engines in one class

The 6ft 2in 2–4–0s worked the main-line goods traffic. The most recent of these were the 29 engines of the 615 class, built between 1874 and 1878. Tractive effort was only 11,611lb, grate area was 15½sq ft and adhesion weight only 25 tons 17cwt, hardly suitable for the heavy main line grades. The 39 0–6–0s were of the same vintage and were the most powerful mineral engines owned by the company. With a wheelbase of only 11ft and an adhesion weight of 37 tons 3½cwt, they were well-suited to sharply-curved colliery sidings, but the outside cylinders and short wheelbase caused unsteadiness at speed and made them unsatisfactory main-line engines.

By 1882 most of the main-line passenger work was in the hands of the 16 Conner 8ft 2in singles, built between 1859 and 1875, and the 38 7ft 2in 2–4–0s, constructed in 1867-1874. The singles were confined to the Carlisle—Glasgow section with one or two runs to Edinburgh. They were unsuited to current requirements and frequently needed pilot assistance. The 7ft 2in 2–4–0s were more useful locomotives but, like the singles, were underboilered by the standards of the middle 1880s. For neither passenger nor goods

THE CALEDONIAN IN 1882

work did the company possess a locomotive comparable in either power or efficiency with those of its competitors.

Another problem facing Drummond was the lack of standardisation. There were no mixed traffic engines. In addition to the 6ft 2in 2–4–0s for goods traffic there were 49 6ft 2in 2–4–0s specifically designed for passenger work. Instead of two or three standard types of boiler suitable for most classes a separate boiler seemed to be designed for each type of locomotive.

The immediate problem of providing more powerful express engines was solved by adapting some of Conner's 6ft 2in 2–4–0s (goods version) for passenger duties. In addition to the provision of a Drummond boiler and chimney, this work involved replacement of the Conner cab with one of Stirling pattern and substitution of closed leading splashers, each combined with a large sandbox, for the original slotted type. A more powerful engine was obtained for very little increase in weight.

	Original	Rebuild
Cylinders, in	18×24	18×24
Boiler: maximum external diameter	4ft 2in	4ft 4½in
Tube heating surface, sq ft	935	838
Firebox heating surface, sq ft	85.5	101
Grate area, sq ft	15¼	17
Boiler pressure, lb/sq in	130	150
Tractive effort at 85 per cent bp	10,718lb	13,398lb
Weight in working order	35tons 17cwt	36tons

The rebuilt engines soon became known as the 'Rebuilds'. The St Rollox boiler-proving register shows their numbers and dates of rebuilding to have been as follows:—

Date	Engine Nos	Date	Engine Nos	Date	Engine Nos.
8/83	420	9/83	427 & 432	11/83	423
2/84	429	4/84	433 & 435	7/84	422 & 431
9/84	428	3/85	430 & 479	4/85	426
5/85	419 & 476	7/85	425	11/85	421 & 483
12/85	480	1/86	424 & 473	2/86	477

For a very short time, the first of these engines handled the best expresses but, with the arrival of the Drummond 4–4–0s, they were relegated to lighter or less important services.

A further six engines were rebuilt with Drummond boilers in 1886 to Order No Y3 though whether or not they were given Stirling cabs is not clear. Their numbers were:—

THE CALEDONIAN IN 1882

Date	Engine Nos	Date	Engine Nos
5/86	475	6/86	478
8/86	380	9/86	241
10/86	288 & 484		

Other Conner 2–4–0s of both the 6ft 2in and 7ft 2in varieties were fitted with the 'rebuild' boiler over the years, some engines getting the Lambie version. The original version was also used in the rebuilding of Brittain's 'Dundee bogies'.

CHAPTER 2

BASIC DRUMMOND DESIGNS

THE 294 CLASS 0-6-0s ('JUMBOS')

The cylinder and motion details of the famous 'Jumbos' were based on Stroudley practice. Pistons and cylinder covers were conical. The steamchests were placed between the cylinders, the mid-length of the valve faces being forward of the mid-length of the cylinders. The steam and exhaust ports in the vertical valve-faces were divided horizontally into equal upper and lower portions separated by a bridge which allowed steam to pass rapidly from one end of the chest to the other. The slide valves extended from top to bottom of the valve faces and covered both portions of the ports. Steam from the lower portion of the exhaust port passed round the outside of the cylinder in a cast passage projecting through the frames and joined that from the upper portion near the base of the blastpipe. This was to prevent the choking effect which might have occurred had all the exhaust been discharged through the limited space between the cylinders.

The slidebars were box-shaped and partially enclosed the crosshead, which was forged solid with the piston rod. Unlike those of the NB 454 class, they were not attached to the cylinder-block but were secured by the motion-plate just behind their mid-length. This meant that support was provided at the point where the greatest stresses fell. The absence of obstruction at the rear of the bars allowed the expansion links to be brought further forward than would otherwise have been possible and the consequent lengthening of the eccentric rods resulted in a smaller increase in lead when the engine was notched up. The connecting rods were rectangular in section but, with their marine big-ends, otherwise

BASIC DRUMMOND DESIGNS

resembled the Stroudley pattern. The crank-axle was solid-forged with parallel-sided webs and the wheel centres were of Krupp cast steel with square-ended balance weights forming an integral part of the casting. The load on the driving wheels was borne by spiral springs of Timmis pattern but laminated springs were used for the other axles. The steam brake cylinder was suspended vertically below the cab and operated the brake-blocks by a system of rods outside the wheels.

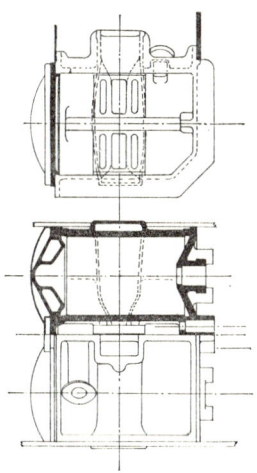

Fig 1   Arrangement of standard 18in × 26in cylinders: 1883

The boiler barrel was composed of two iron rings arranged telescopically, the rearmost being the larger. The dome carried a pair of Ramsbottom safety-valves and contained Drummond's standard regulator, which was of the vertical sliding or grid-iron type with a small pilot valve at the back for ease of operation. There were 216 brass tubes of 1¾in diameter and 10 of 1½in diameter. Boiler pressure was 150lb/sq in. Five curved wrought-iron bars were rivetted to the underside of the outer firebox casing and from each of these was slung a cast steel bracket which formed an anchorage for three stay-bolts uniting with the elliptical roof of the inner firebox. The boiler was fed by two No 8 brass injectors placed outside the frames opposite the ashpan. These were designed and made at St Rollox and were known as 'Drummond' injectors. The clack-valves were mounted on the front ring of the

BASIC DRUMMOND DESIGNS

boiler-barrel. The ashpan was provided with cast iron double doors, front and rear, which were machined at the meeting surfaces to provide air-tight conditions when shut.

The orifice of the plain blastpipe was at a level slightly below the top row of tubes and was encircled by the blower ring. There was no petticoat-pipe or chimney liner. The chimney itself was composed of an upper and a lower plate, butt-jointed to give a smooth contour. The lower provided the base and lower 8in of the barrel while the upper carried the cast iron top, 10in deep. The smokebox door had two plain handles. The main hand-rails alongside the boiler terminated half-way along the sides of the smokebox, the left-hand rail carrying the control rod from the cab to the blower valve. A separate rail followed the upper circumference of the smokebox door, and below and on either side of the door was a small, square, brass tallow-cup, with T-shaped tap, for cylinder lubrication.

The faceplate, lagged to keep the cab cool, carried the regulator handle, two water-level gauges, the regulator lubricator (left-hand side) and the steam brake valve (right-hand side). The Stroudley-type firedoor consisted of a cast-iron flap, hinged at the top, and operated by a lever and notched quadrant. The steam valves for the injectors were controlled by two solid brass wheels on rods projecting through the spectacle-plate above the firebox. Between these was the whistle-rod, while below the left spectacle-glass was the T-shaped blower handle. A wrought-iron box with a teak lid occupied either side of the cab, the one on the left carrying a sector with levers controlling the cylinder-cocks and the gravity sanding. The reversing lever was also on the left, its quadrant having six notches on either side of mid-gear. Projecting through the floorboards were the handles operating the water-valves to the injectors and, to the right of the reverser, the levers working the dampers.

The double-framed tender was almost pure Stroudley design. The outer frames had long oval slots but the inner frames formed the outer walls of a well. The tank was shorter than the frames and was clothed in steel sheeting, $\frac{1}{8}$in thick, which left an air-space of $\frac{3}{8}$in between the tank-plates and the clothing. A toolbox rested on the platform above the frames at the rear of the tank. The brakes were operated by hand only and the springs were underhung. Two versions of this tender were built for the class, the only difference being the length of the well.

BASIC DRUMMOND DESIGNS

The water capacity of the 2840-gallon pattern tender was sometimes given as 2800 gallons. This was simply a round figure; the weight of water in the so-called 2800-gallon tender corresponded to 2840 gallons.

Because of the pending reorganisation of the works, the engines were put out to private contract. On 26 December 1882 the board opened tenders for the supply of "Ten Passenger Engines and Tenders and Fifteen Goods Engines and Tenders", and on the following day Neilson's offer was accepted, the 25 engines being priced at £2900 each. According to the contract (Neilson order number E561), delivery was to start in 10 months and be completed within a year. In fact, the first of the 15 0–6–0s arrived at St Rollox on November 13 1883 and the last on January 19 1884. They carried the Neilson works numbers 3043-3057 and were given the CR running numbers 294-308. They and certain other early batches of the class had iron crank-axles and 2500-gallon tenders.

On 28 March 1883 Drummond recommended that 10 goods engines be built in the works and that work on five be begun immediately. In fact, *six* engines, Nos 349-354, were completed to this order between November 1883 and January 1884 at a cost of £2000 each. The last three differed from the others in having steel crank-axles. All six had 2500-gallon tenders. A further 20 engines were ordered from Neilson & Co. in December 1883, again at a cost of £2900 each. (Neilson order No E580, 12 December.) The 20 engines were delivered between July and December 1884 and, though carrying consecutive Neilson works numbers, were numbered in the Caledonian list as two separate batches, Nos 517-526 (Neilson 3252-61) and 680-689 (Neilson 3262-71). All had iron crank-axles and 2800-gallon tenders. The chimney cap of the original batch had been like that of the NB engines but this now gave way to a new pattern with a shallower cap. All engines were painted dark blue.

By the time the last of the Neilson engines was at work, St Rollox was able to assume responsibility for all new locomotive construction. When more goods engines were required, four further batches of 0–6–0s were laid down:—

BASIC DRUMMOND DESIGNS

Order No	Date ordered	Engine Nos	Dates delivered	Cost per engine
SW	4/85	690-5	12/85	£2000
Y5	11/85	309-20	6/86—9/86	£2000
Y9	5/86	355-60 366-71	12/86—2/87	£1700
Y12	10/86	321-2 339-48	5/87—8/87	£1700

Engines of the first two batches had steel crank-axles but a reversion to iron was made for lots Y9 and Y12. All tenders were of 2840 gallons capacity. No 355 was given a spare boiler built to order No Y5 and, possibly because of that, was recorded in the locomotive register as delivered in October 1886.

Of the early batches, two engines went to Aberdeen, two to Perth and two to Stirling, while Carlisle got Nos 339 and 367 amongst its rather larger allocation. The working timetable of April 1886 gave their duties as follows: Carlisle, Perth, Glasgow and Greenock: Dundee and Greenock: Aberdeen, Perth, Glasgow and Edinburgh: Edinburgh, Aberdeen and Arbroath. On Beattock Bank they were allowed to take up to 25 wagons without assistance.

THE 66 CLASS 4–4–0s

Much of the new express engine designed at the same time as the Jumbos, including the cylinders and the boiler, was identical to that of the 0–6–0s. The frames were 4ft 1½in apart except for a short length opposite the front bogie-wheels where they were stepped in by 1in on either side. Semicircular notches were cut out of the frames behind the bogie-wheels to provide clearance on curves. The bogie was of the Adams pattern with pivot equidistant from the axles. The load on the driving wheels was borne by springs of Timmis pattern, with laminated springs for the trailing axle. The slidebars were attached to the cylinders in front and were supported at the rear by the motion-plate, the position of which limited the length of the eccentric rods to 4ft 7in. The weighshaft and balance-weights were below the motion, whereas in the 0–6–0s they were above it.

The Westinghouse brake cylinders were placed vertically between the coupled wheels and applied the brake-blocks to the rear of the driving wheels and the front of the trailing wheels. The exhaust pipe from the Westinghouse pump ran alongside the outside of the

BASIC DRUMMOND DESIGNS

boiler, beside the handrail, before entering the smokebox. The steam cock operating the pump occupied the same position on the faceplate as the steam brake cock of the 0–6–0s. Reverse was by lever, the quadrant having eight notches on either side of mid-gear. The levers operating the sanding gear and the cylinder cocks were mounted in separate quadrants on the left and right hand sides of the footplate, respectively.

The first of the 4–4–0s arrived at St Rollox on 22 February 1884 and the last on 12 April 1884. The Neilson works numbers were 3058-3067 and the CR running numbers 66-75. A further batch of six engines was ordered in April 1884, this time from St Rollox (order number Y). They were turned out between February and June 1885 at a cost of £2234 each and were numbered 60-65. Both batches had steel crank-axles.

The tenders were identical to those of the goods engines, except for the addition of the Westinghouse brake. Tender drawings for the Neilson batch show the water capacity to have been 2500 gallons. This figure is confirmed by the Neilson diagram book which gives the weight in working order (with four tons of coal) as 31 tons 9cwt. The locomotive register, however, gives the following water capacities:—

Nos 69 and 72	2500 gallons
Nos 60-65	2800 gallons
Nos 66-68, 70, 71 & 73-75	2840 gallons

Again it seems probable that the 2800- and 2840-gallons tenders were identical.

Most of the class were allocated to Perth and Carlisle. The remainder were distributed between Polmadie, Aberdeen (two engines) and Stirling (one engine). Their arrival coincided with a general improvement in the express services:—

Year	Number of trains averaging over 40 mph	Daily mileage	Average running speed, mph
1883	16	1156	42.75
1888	46	2989	43.2

Most of these accelerations applied to the Anglo-Scottish traffic.

BASIC DRUMMOND DESIGNS

The average running speeds were still low and the weights of the trains differed little from those of the late 1870s but the greater power of the new engines made itself felt by eliminating much of the double-heading, which at that time resulted in high pilot engine mileage. The down Limited Mail, before the advent of the Drummond locomotives, was piloted regularly from Carlisle to Beattock Summit and sometimes even as far as Carstairs. The success of Drummond's 'big engine' policy is seen in the official analysis of locomotive expenditure for 1887:—

> At January 1883 we had 69 Pilot Engines working but this number has been reduced to 24 and further reductions are being made. This decrease is caused by more powerful engines being placed on the Traffic and the assisting and light running mileage has decreased from 882,680 to 274,990.

From the operating point of view, this was the outstanding contribution made by the Drummond 4-4-0s during the years 1884-90. Opportunity for fast running was limited and it was not until the 1890s that the engines were really given their heads.

EARLY TANK ENGINES

Drummond's first Caledonian tank engines were the 171 class 0-4-4Ts. They were derived from the 72 class 4-4-0Ts of the North British, the dimensions of the cylinders, wheels and boiler-barrel being the same. They were intended for light branch-line duties but resembled the larger Drummond engines in most points of design.

The ports were undivided, the whole of the exhaust steam passing direct to the blastpipe. The slidebars and the weighshaft were arranged as in the 'Jumbos'. The boiler was fed by two No 8 injectors but the steam cocks for these were positioned on the face-plate, that on the right side also supplying steam to the Westinghouse pump. Reverse was by lever, the quadrant having five notches on either side of mid-gear. The levers operating the cylinder cocks and the sanding gear (fitted for forward running only) were situated on top of a box in the left rear corner of the cab. The brake valve was attached to the side-sheet *behind* the entrance. Other unusual features were the level grate, the solid bogie-wheels and laminated springs for the crank-axle. Water capacity was 830 gallons and coal 1¼ tons.

BASIC DRUMMOND DESIGNS

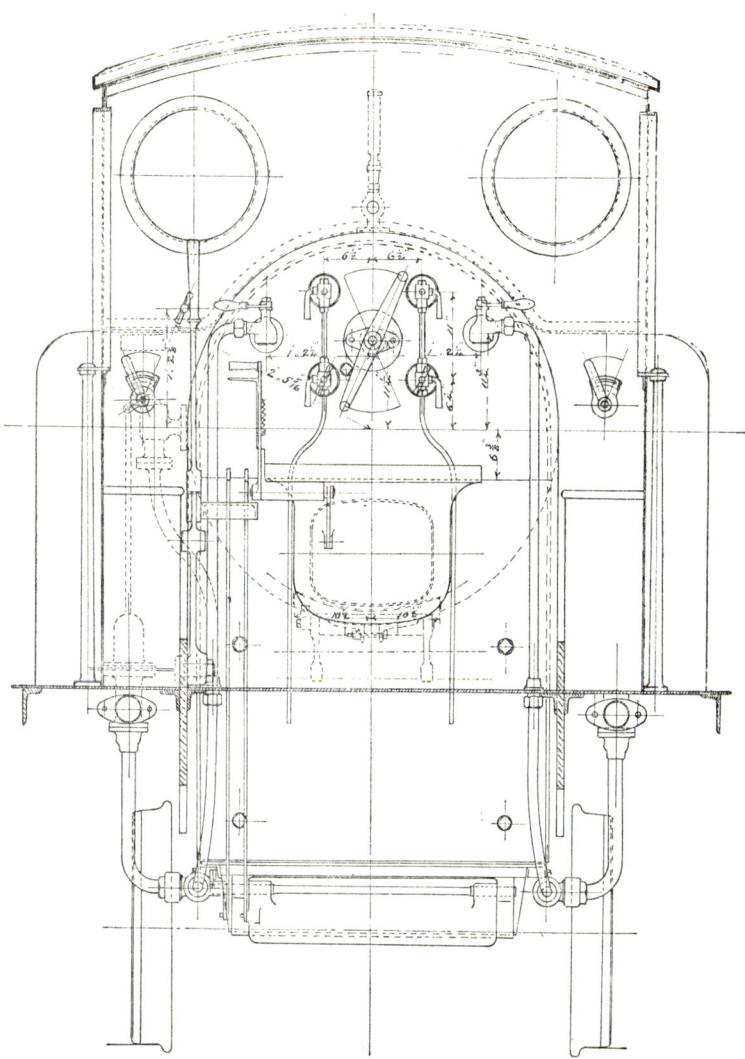

Fig 2 Footplate layout of 171 class as originally designed without Westinghouse brake

BASIC DRUMMOND DESIGNS

The order was for 12 engines.

Order No	Date ordered	Engine Nos	Dates completed	Cost per engine
SN	11/83	171-6	5/84—8/84	£1398
SN second	4/85	177-8 228-31	2/86—5/86	£1398

In Nos 171-6, the crankaxle and frames were made of steel. The class was used for light work on country branches, and it handled the Balerno branch until 1899.

Following the first batch of 0–4–4Ts came 10 outside-cylinder saddle-tanks. The first reference to these engines is in a minute dated 2 December 1884, which invited Drummond to make drawings and submit an estimate of the cost of construction. The matter was further considered by the locomotive and stores committee on 16 December:—

> Pug engines for working mineral traffic at Stobcross and General Terminus Docks.
> Submitted specification for 10 pug engines—estimated cost £800 each. Mr Drummond produced plans and stated that the cost of one will be £800 but if the whole are sanctioned the price will be £700 each. Decision to be given at next Board.

When the board met on 30 December, the potential £100 saving on each engine was apparently overlooked and the opportunity was therefore taken to complete two of them, Nos 262-3, as 0–4–2STs as a cost of £975 each and the remainder, Nos 264-71, as 0–4–0STs at £720 each. All ten engines were delivered between June and September 1885 to order No Y1.

These engines were derived from the standard Neilson 0–4–0STs of 1875, of which 14 were already owned by the company. In building further examples, Drummond adopted the boiler, cylinders, motion and wheels of the Neilson engines but retained his own chimney, regulator, smokebox and cab details for those of the original design. Features foreign to Drummond practice included the three-ringed boiler-barrel, the flat-topped inner firebox stayed by longitudinal girders, the cast-iron wheels with eight spokes of T-section and the reversing lever on the *right* hand side of the footplate.

The 0–4–2STs had full-length cabs and proper bunkers with a large toolbox at the rear, as in the 0–4–4Ts. Although the rear

BASIC DRUMMOND DESIGNS

axle had sideplay, the engines' range of action was limited to a maximum curvature of 4½ chains; the 0–4–0STs could negotiate curves of 1½ chains radius. The two engines were hand braked but the Westinghouse brake was added about 1887 as a Board of Trade condition for their working the steeply-graded Killin Railway. Both engines were painted blue and operated the Killin branch until replaced by two 171 class 0–4–4Ts in 1895.

Dimensions of the 0–4–2STs were:—

Total wheelbase	13ft 9in
Trailing wheel diameter	2ft 6in
Length over buffers	27ft 8¼in
Coal capacity	1¼ tons
Adhesive weight	25 tons 17cwt
Total weight	31 tons 4½cwt

The 0–4–0STs built to order Y1 had no rear bunkers. Instead, the lower half of each cab side-sheet was extended to form a small coal store on either side of the firebox. Sometimes coal was carried on the already meagre footplate. Later, small wooden trucks carrying about 1½ tons of coal were supplied. The engines were painted black and had dumb buffers. The reversing quadrant had six notches on either side of mid-gear, though there could have been little opportunity for expansive working! Water capacity was 800 gallons and coal 11cwt.

PERSPECTIVE

By the end of 1885, Drummond had constructed 63 main-line engines—47 0–6–0s and 16 4–4–0s. In addition, eight tank engines had been built for light passenger work and a further eight for dockyard duties, while 21 of the Conner 2–4–0s had been rebuilt with large boilers. These engines not only satisfied the company's need for motive-power but also provided a solid foundation on which further locomotive development could be based. The company now possessed engines as advanced as any in the country. In a report on locomotive expenditure, dated 24 October 1887, Drummond showed the extent to which he had been able to reduce costs.

			Pence per Train Mile		
Month	CR	NB	LNWR	NE	Midland
1/83	12.28	9.27	9.65	14.90	9.97
7/87	10.56	8.46	10.71	15.63	11.77

The two Scottish companies alone showed a continuous decrease. The number of engines in steam daily at the time of the report was 601 for the Caledonian and 587 for the NB. For this increase of 14 engines, 162 fewer men were required:

Company	Drivers	Firemen	Cleaners
NB	663	673	424
CR	589	626	383

In summing up the improvements he had made Drummond concluded:—

> The condition of the Rolling Stock Five Years ago and now; and the difficulty in carrying on the necessary Repairs during the Rebuilding and Rearranging of the workshops must be taken into consideration; and not withstanding this difficulty, we have now been able as shewn each Half Year to reduce our Expenditure below the cost per train-mile of the best equipped line in England.

No wonder the board had increased his salary to £2000 per annum in October 1884.

CHAPTER 3

IMPROVING THE BREED

EXHIBITION ENGINES

At the 1886 Edinburgh International Exhibition two new Caledonian express engines were on show, No 123 built by Neilson & Co. and No 124 by Dübs & Co. It is not clear whether the builders or the railway company first proposed the construction of a special locomotive for the exhibition. No 123 seems to have been conceived solely by her builders. In a letter to *Engineering*, Edward Snowball, Neilson's chief draughtsman, wrote:

> The general design and the whole of the detail drawings were made in Messrs Neilson's drawing office under the immediate supervision of the head of the firm, without any intervention on the part of the Caledonian locomotive superintendent.

Neither Drummond nor anyone else in authority challenged this statement and it was certainly unlike Drummond to let someone else take the credit for work which was his own.

In the Neilson sketch book there is an outline drawing with leading dimensions dated 21 December 1885, in which the engine looks like a Brittain version of a Conner 8ft single with leading bogie and Drummond boiler, boiler-mountings, cab and tender. The driving wheels were 8ft 3in in diameter and the outside cylinders, 19in by 27in, projected through a gap in the running-plate, as in George Brittain's Oban bogies. This design was not one to appeal to Drummond and nothing came of it.

On 29 December 1885, eight days after the above project had been committed to paper, the following CR board minute was recorded:—

IMPROVING THE BREED

New Engines. Opened tenders for two New Engines. Accept Messrs. Neilson & Co's offer for one at £2600 and Messrs. Dübs & Co. for one at £2600 for Exhibition purposes.

It seems that Neilson had still to decide on the exact form its engine was to take. The official offer was not made until 22 January, and the engine then proposed was a 4–2–2 which, tender apart, was fully in the Drummond style. The Neilson order (No E600 of 23 January) ran:—

The Caledonian Railway Company.
One Bogie Express Passenger Engine & Tender to drawings and specifications our offer of 22nd inst. and their acceptance of date. To be exhibited at the Edinburgh International Exhibition where it must be delivered by 1st April 1886 without fail.

The cylinders, slide valves, motion and reversing gear were modelled directly on those of the 66 class 4–4–0s. The slidebars were attached to the cylinders in front and were supported by the motion plate at the rear. The weighshaft, balance weights and reach-rod were below the motion and the reversing quadrant had eight notches on either side of mid-gear. The boiler barrel was of the same length as that of the 4–4–0s but was $2\frac{1}{8}$in smaller in diameter and pitched 3in higher. The firebox was $8\frac{3}{8}$in shorter and the tubes 30 fewer. The boiler was fed by two No 8 brass injectors and the clack-valves and boiler mountings were standard Drummond type; the chimney, however, was slightly reduced in height to compensate for the higher-pitched boiler.

Although closely modelling their engine on Drummond practice, the builders showed their capacity for innovation. Adequate adhesion was provided by fitting the system of compressed air sanding invented by the Derby works manager, F. Holt, in 1885. This employed a jet of air from the Westinghouse brake system. The frames were made of steel instead of Yorkshire iron and the blastpipe was of the vortex type introduced by William Adams on the LSWR in 1885. This was composed of an outer annular orifice for the discharge of exhaust steam, and an inner pipe, the lower end of which curved backwards and opened out to collect exhaust gases from the lower rows of tubes. The annular form of the exhaust steam orifice provided a jet of greater surface area for entraining the exhaust gases than could be achieved with the conventional blastpipe. Twin organ-pipe whistles were fitted, the small

IMPROVING THE BREED

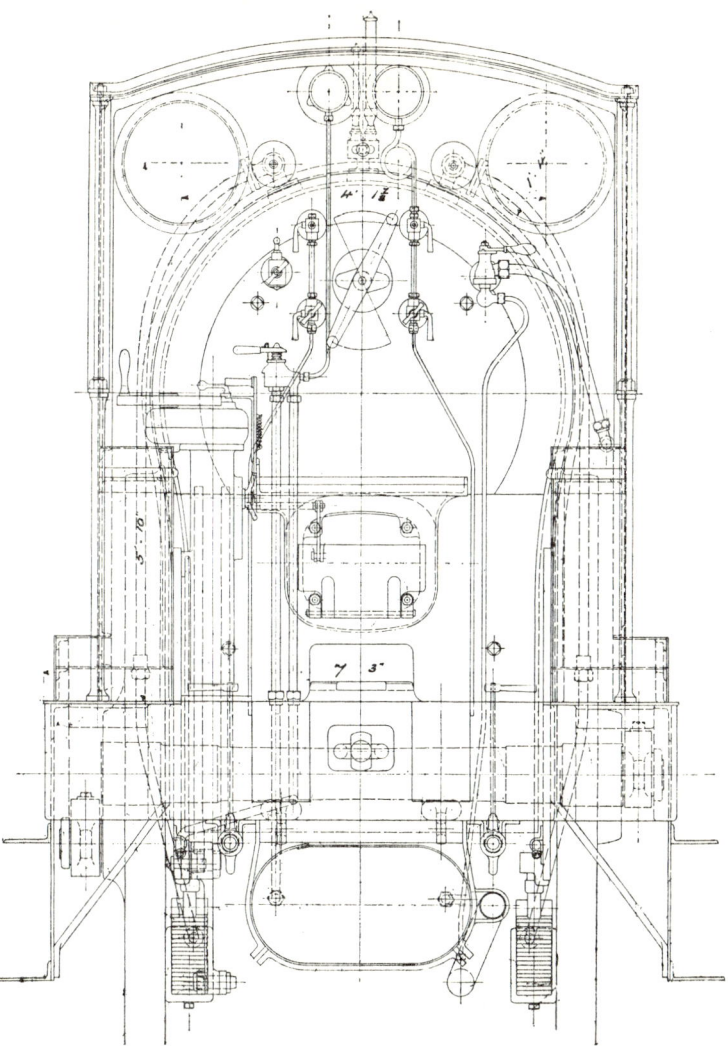

Fig 3 Footplate layout of No 124. Note vertical screw reverser

IMPROVING THE BREED

one on the left being intended for use in stations and yards. A continuous splasher covered the bogie-wheels on either side. The pipe carrying the exhaust steam from the Westinghouse pump passed vertically downwards and then ran along the underside of the running plate before entering the exhaust passage on the outside of the cylinder.

The original Neilson sketch of December 1885 shows that the tender then contemplated for No 123 was of the standard Drummond type. The new tender had outside frames with arc-shaped slots. The springs were placed above the axleboxes and the coal-flare was radiused outwards from the tank. A well of the standard breadth and depth but 14ft 0½in long was placed below the tank.

No 123 was given a special finish for the exhibition. The axle-ends were polished bright and a new alloy, Arguzoid, which took on a bright, silvery appearance when polished, was used for the chimney cap and the cab fittings. The Caledonian coat-of-arms appeared on the sides of the tender but the company's initials were omitted. Brass works plates carrying the number 3553 were fixed to the centre of the splashers. Engine and tender panels are said to have been lined out in gold.

The Dübs engine, No 124, was basically a 66 class 4–4–0 with a completely redesigned front-end, Bryce-Douglas valve gear and a vertical screw reverser similar to that already used by Drummond on the NBR. Archibald Bryce Douglas was a senior engineer in the Fairfield Shipbuilding and Engineering Co. of Govan and was a friend of Drummond's. His valve gear was intended to improve steam distribution in slow-moving marine engines but certain features made it attractive for locomotives; eccentrics were not required and the layout of the gear allowed the slide valves to be placed above the cylinders without the need for rocking levers. The absence of eccentrics allowed the provision of wide journals (8¼in compared to 7½in in the 66 class) and thick crank webs (5¼in as opposed to the previous 4⅝in) in conjunction with 19in cylinders without restriction of the space available for the discharge of exhaust steam. As with Walschaerts gear, part of the movement imparted to the valve spindle was derived from the crosshead, but transmission was indirect, the crosshead being coupled not to the valve spindle but to the expansion link. In addition to the reciprocating movement provided by the crosshead, the expansion link was given an oscillatory motion by means of a complicated

Page 33 (upper) Drummond 0–6–0 No 306; *(lower)* Drummond 0–4–4T No 1223 at Perth

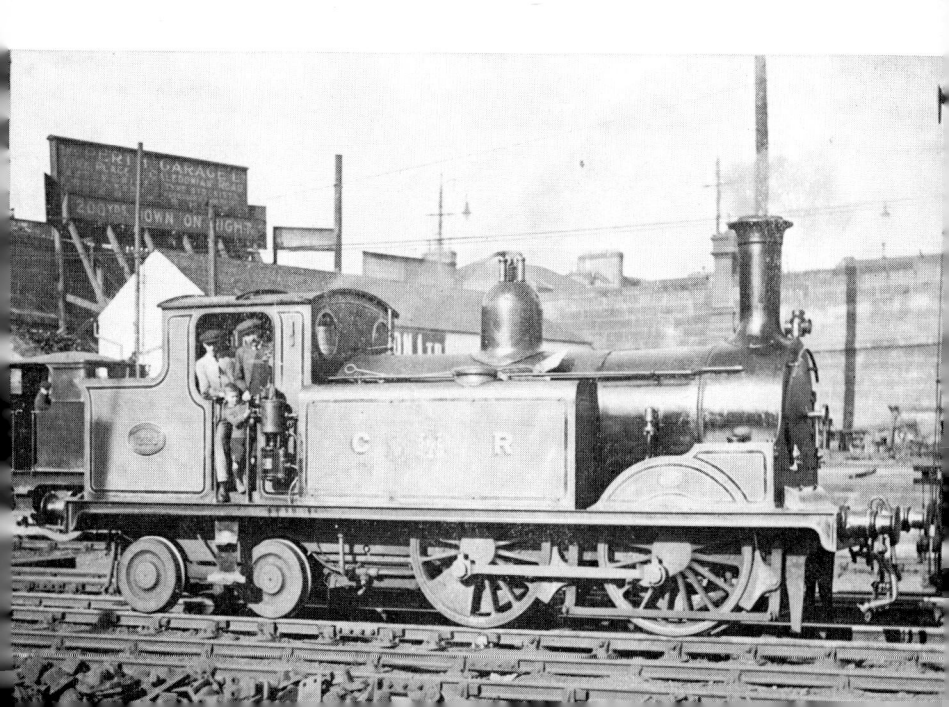

Page 34 (*upper*) Drummond 4–4–0 No 90 with Drummond 3130 gallon tender at Glasgow Central; (*lower*) 'Dunalastair' No 723

IMPROVING THE BREED

arrangement of links which put it in communion with the connecting-rod.

The cylinders and valve chests were cast in one piece. The ports were of the same dimensions as those of the 66 class but, because of their position on top of the cylinders, were undivided. The slide-bars and motion plate were arranged as in the existing 4–4–0s. Below the smokebox, the frames were extended upwards to cover the valve chests and this necessitated an increase of $1\frac{1}{2}$in in the pitch of the boiler, the chimney being correspondingly reduced in height to keep the engine within the loading gauge. In all other respects, the boiler was identical to that of the 66 class. No 124 had gravity sanding, a plain blastpipe, frames of Yorkshire iron and a standard double-framed tender of 2840 gallons capacity. She had a double whistle and was fitted with train-heating equipment. A passage in the cylinder-block casting collected exhaust steam from the base of the blastpipe and passed it through a pipe which ran back below the right hand side of the engine. The exhaust pipe from the Westinghouse pump and live steam from the Westinghouse steam cock on the faceplate also discharged into this. Below the footplate was a valve controlling the amount of steam passed to the train, which was operated by a vertical rod extending into the cab.

No 124 was finished in standard blue but her tyres, wheel centres and spokes were intricately lined and the axle-ends were finished bright. The tender carried the company's initials but without the coat-of-arms. The diamond-shaped Dübs plate bearing the works number 2245 occupied the driving wheel splashers and was surrounded by decorative thistle leaves.

Both engines were awarded gold medals and were acclaimed by the technical press. After the exhibition they went through St Rollox Works before entering service in December 1886. No 123 was sent to Edinburgh and was employed on the principal London expresses, notably the 10 am. She was still so employed at the time of the 'Race to Edinburgh' and was the only engine used by the Caledonian on the race train. Her average time for the 100.6 miles from Carlisle to Edinburgh during August 1888 was $107\frac{3}{4}$min, an average speed of 56mph. Her fastest time was made on 9 August when, with a load of four eight-wheeled carriages weighing approximately 80 tons tare, Edinburgh was reached in 102min 33sec, the highlight being the average speed of 56.9mph over the

IMPROVING THE BREED

41.1 miles from Gretna Junction to Beattock Summit, where the mean up-gradient is 1 in 207. She remained at Edinburgh for several years and was employed mainly on the 10 am London express and the corresponding down train from Carlisle. Around 1889, she was given a Drummond 3130-gallon tender (see later), her original tender being acquired by 'Jumbo' No 413.

No 124 was sent to Polmadie. Her early performance was marred by trouble with her complex valve gear which caused three total failures within six months. The decision to rebuild her with standard 18in by 26in cylinders and Stephenson link motion was taken in June 1887. The frames were not altered; their increased depth below the smokebox (compared with those of the 66 class) remained a distinguishing feature.

For some time after conversion, No 124 was the star express engine on the line. She went through St Rollox works in May 1888 in preparation for royal train duty between Motherwell and Glasgow on the occasion of the opening of the Glasgow International Exhibition. The ceremony was performed by the Prince and Princess of Wales and No 124's tender was appropriately adorned with the heraldic feathers of the Prince of Wales. By then, however, her original tender had been replaced by one of the new 3130-gallon tenders (see p.39).

In the late 1880s No 124 earned a high reputation on the Carlisle main line and in 1893 J. Pearson Pattinson described several notable runs behind her. The most detailed of these—between Carlisle and Carstairs—is set out in Table 1 alongside No 123's record run in the Race to Edinburgh. The 4–4–0 had a much heavier train—'equal to 15' or about 204 tons tare—but made a very fast start, the first mile being covered at an average speed of 45mph. She was only four minutes behind the single-wheeler's time at Beattock but fell behind on the bank proper. The 35.6 miles from Rockcliffe to Beattock were covered at an average speed of 56mph. On Beattock bank the engine probably just managed to sustain a speed of 27mph around milepost 46, for the 46th and 47th miles were covered in precisely the same time and the 48th took only one second longer. Pattinson commented, 'This is a grand performance and we doubt if it has ever been beaten'.

In 1890, No 124 was transferred to Ardrossan to work The Arran Express boat train, an eight-coach formation specially built for the opening of the new route from Glasgow to Arran via

IMPROVING THE BREED

TABLE 1
EARLY PERFORMANCE

Engine No Load	Distance (miles)	123 =6 Time		124 =15	
		min	sec	min	sec
Carlisle	0.0	0	00	0	00
	4.0			5	01
Rockcliffe	4.1	5	35		
Gretna Jtc	8.6	9	38		
	9.0			9	58
	13.0			14	43
Kirkpatrick	13.1	14	07		
Kirtlebridge	16.7	17	46		
	17.0			19	25
	20.0			22	39
Ecclefechan	20.1	21	04		
Lockerbie	25.8	26	46	28	51
Nethercleugh	28.7	29	22		
	29.0			31	49
Dinwoodie	31.7	31	54		
	32.0			34	42
	34.0			36	50
Wamphray	34.5	34	26		
	35.0			37	52
Beattock	39.7	39	13	43	13
	42.0	41	44	46	42
	44.0	44	20	50	35
	46.0	47	17	54	57
	48.0	50	23	59	24
	49.0	51	58	61	39
Beattock Summit	49.7	53	04	63	19
Strawfrank Jtc	73.2	74	44		
Carstairs	73.5			86	38

Ardrossan. She was named *Eglinton* in honour of the Earl of Eglinton on whose estates the town of Ardrossan was built. She continued on this work until about 1906-7 when she was moved to Perth for a few months and then returned to Polmadie.

THE 'COAST BOGIES'

Drummond's next design was for a small-wheeled 4–4–0 intended chiefly for the services to the Clyde coast but suitable also for general excursion traffic. Because of their association with the Gourock line, these engines were known as the 'Coast Bogies' or 'Gourock Bogies'.

IMPROVING THE BREED

The slidebars and motion plate were arranged as in the Jumbos, the cylinders and the whole of the inside motion being interchangeable between the two classes. The crank axle was made of iron and reverse was by lever, with eight notches on either side of mid-gear. The bogie, suspension and brake arrangement were virtually identical to those of the 66 class but the exhaust pipe from the Westinghouse pump ran vertically down to join the train-heating pipe below the ashpan.

The boiler barrel was the same length as that of the 66 class but, although pitched at the same height, was $2\frac{1}{8}$in less in diameter. The coupled wheelbase was 1ft shorter than that of the express engines and this restricted firebox length and grate area, this being only 16.75sq ft. The tubes were of brass and the boiler barrel of Yorkshire iron.

The design of the 'Coast Bogies' was influenced by that of the two exhibition engines. Features borrowed from No 123 included the vortex blastpipe (though of modified form), steel frames and compressed air sanding. Twin organ-type whistles and train-heating equipment were fitted, the steam being collected from the exhaust passage on the outer face of the right-hand cylinder. The pipe arrangement at the rear of the engine was similar to that described for No 124.

The tenders were standard, with underhung springs and toolbox at the rear. Water capacity was recorded by *The Engineer* as 2840 gallons and by the locomotive register as 2500 gallons. These tenders were replaced around 1890, the replacements having the same tank and toolbox arrangement as the originals (water capacity being 2840 gallons) but deeper frames with springs above the axleboxes.

The first six engines of the class were ordered in January 1887:—

Order number	Engine Nos	Dates delivered	Cost per engine
Y13	80-2, 85-6, 116	2/88—4/88	£1667

They were employed on the Glasgow-Greenock route, four being allocated to Polmadie and two to Greenock. On 27 April 1889, one of these engines (probably No 80) was used to haul a special train of newly introduced stock conveying officers and pressmen from Glasgow to Loch Awe and back. The other five engines remained on the coastal traffic until about 1917.

IMPROVING THE BREED

TENDERS

Before 1888, both pasenger and goods engines were fitted with the Stroudley-type tender of 2500 or 2840 gallons capacity. Work on the design of a new tender began in November 1886, the principal innovation being the placing of the springs above a new type of axlebox. The outside frames were brought closer together than in the Stroudley pattern but were of the same depth, only 2ft between horns. The inside frames again formed the sides of a well which was identical in dimensions to that of the 2840-gallon tender. The tank was slightly narrower but deeper than that of the earlier tenders. Water capacity was 3130 gallons, the increase being achieved by transferring the toolbox to the top of the tank and extending the tank to the rear of the frames.

It has been stated that the Stroudley-type tenders rode badly and gave trouble with overheated axleboxes and journals and that for these reasons underhung springs were abandoned. There is no indication that similar trouble was experienced with the Drummond tenders on the NBR. A statement on the tender heating problem (*The Locomotive*, 1902) refers not to the Stroudley-type tender but to the Drummond 3540-gallon pattern of 1889. A more likely problem with the early tenders was insufficient water capacity; extension of the tank to the rear of the frames was a first step in its solution.

According to the locomotive register, the only engines given the 3130-gallon tender were the second batch of 'Coast Bogies' (Y28) and No 116 of the first batch. Photographic evidence, however, proves that the express 4–4–0s of 1884-6 and No 123 also acquired tenders of this type. The change of tenders probably occurred in 1888 for No 124 had one of the new type when engaged on royal train duties that summer. The originals were kept for use with future 'Jumbos'.

New tenders carrying 3550 gallons of water (probably a round figure since capacity was sometimes quoted as 3540 or 3560 gallons) were built for the Y21 series 4–4–0s of 1889, these engines being intended for through working between Carlisle and Aberdeen. The hornblocks, axleboxes and springs were modelled on those of the 3130-gallon tender but the outside frames were of increased depth and the slots between the hornblocks were arranged in pairs. The wheelbase was the same as that of the earlier tenders but the rear overhang was lengthened. This allowed the tank and the well

IMPROVING THE BREED

to be extended. The toolbox was placed on top of the rear end of the tank which was 3in deeper than that of the 3130-gallon tender.

From the beginning of 1888 the lubrication of tender journals was improved by the adoption of the 'metallic lubricator' which was fitted to some, if not all, of the 3130- and 3550-gallon tenders built in 1888-9. It consisted of an arrangement of pulleys and springs at the bottom of the axlebox, the pulleys being sprung against the lower surface of the journal. As the journal rotated, the pulleys revolved with it, oil being carried to the journal from the bottom of the axlebox. The number and layout of the pulleys was changed several times during 1888-9 and extra springs were added but success seems to have been elusive and from February 1890 the pulleys were replaced by a pad held against the lower surface of the journal and supplied with oil by conventional tail-trimmings.

Drummond's last tender was designed in the autumn of 1889 for the Y23 series Jumbos and was a modernised version of the old 2500-gallon pattern with deeper outside frames and springs above the axleboxes. A 2840-gallon version was attached to the Y13 series 'Coast Bogies' in 1890-1. Like the 3130- and 3550-gallon patterns of 1888-9, these tenders had hook-shaped spring-hangers.

TANK ENGINES 1887-1890

Before 1887 most shunting duties were performed by four-coupled tender engines, the company owning only 23 0–6–0Ts. To provide more suitable power, Drummond designed an 0–6–0 saddle-tank with 18in by 26in inside cylinders and the 'rebuild' boiler, this getting its name from the fact that it was almost identical to that used in the rebuilding of the Conner 2–4–0s. The footplate was open, the cab being of Stirling pattern, but the back-plate of the bunker was extended upwards to form a rear weatherboard. The saddle-tank covered the full length of the boiler and firebox and abutted against the front of the cab. The bottom of the tank was level with the centre-line of the boiler, this allowing the clack-valves to be fitted to the sides of the front ring in the usual way. The boiler was fed by two No 8 injectors of the standard type, placed above the running plate, opposite the firebox. Steam supply to the injectors was taken from two cocks on the faceplate, as in the 171 class.

The wheel-centres were of cast-iron with 10 spokes of T-section.

IMPROVING THE BREED

Sanding was by gravity, the front sandboxes being integral with the leading splashers and the rear ones attached to the exterior of the frames, below the footplate. The levers controlling the cylinder-cocks and the front and rear sanding were located on top of a wrought-iron box on the left side of the cab. The steam brake cylinder was slung vertically below the footplate and operated the brake-blocks by outside brake-rods. The driver's brake-valve was on the faceplate to the right of the regulator, but the reversing lever was on the left-hand side of the footplate. The smokebox, blower and lubricator details were pure Drummond but the buffers were of Conner pattern. Water capacity was 1000 gallons and coal $2\frac{1}{4}$ tons.

Four batches of the class were built:—

Order No	Date ordered	Engine Nos	Dates delivered	Cost per engine
Y14	1/87	323, 385, 506-7, 538-9	8/87—9/87	£1200
Y18	9/87	216-7, 232-5	12/87—1/88	£1200
Y20	7/88	218-21, 386-7	11/88—12/88	£1190
Y24	8/89	388-9, 391-2, 394-6, 398-402	5/90—8/90	£1295

No 323 was renumbered 505 in 1895. In the last three batches, the distance between the driving and trailing axles was increased by 9in, the cab being lengthened and the rear splasher omitted. The crank axles of Nos 391-2, 394-6 and 399-402 were of steel whereas those of the remainder were of iron.

The 0-6-0STs were painted black and spent most of their time in goods yards. The majority were distributed between St Rollox, Polmadie and Edinburgh but Motherwell and later Grangemouth each got a number. Two were at Perth and, in 1902, No 391 was at Aberdeen. Two of the St Rollox engines, often Nos 400 and 402, could be found in Buchanan Street Goods Station, from which they ran trips to Robroyston yard, while another shunted the wagon yard at St Rollox. Known to the enginemen as the 'Jubilee Pugs', they were one of the few Caledonian classes to acquire a nickname.

Further outside-cylinder saddle-tanks were added to stock. The first two were ordered not by the Caledonian but by its joint-line subsidiary, the Glasgow & Paisley Joint Line Committee. They were built by Neilson & Co. in 1887 to order No E607 of 3 December 1886. The order-book contained the instructions 'Delivery within 13 weeks from date. Old 14in drawings to be used where possible'. The two engines carried the Joint Line numbers 1 and 2, on a standard Drummond numberplate, and the Neilson works numbers 3616-7. They were similar to the Caledonian engines of batch Y1 but worked at 150lb/sq in and differed slightly in weight. The No 8 Drummond injectors and the $\frac{7}{8}$in thick steel frames would have sufficed for an engine twice the size. Another feature was the steel inner firebox. This may have been new to the class, for the boiler-proving register shows that at least two engines of batch Y1 originally had copper fireboxes. The Joint Line engines were painted crimson lined out in the Caledonian *passenger* style. Although both came to St Rollox for repairs, responsibility for one of them was accepted by the GSWR. Until 1919 the Caledonian included one in its returns but in that year the GSWR took control of both engines.

In 1888 a six-coupled version of the standard 0–4–0ST emerged from St Rollox:—

Order No	Date ordered	Engine Nos	Dates delivered	Cost per engine
Y16	5/87	272-4, 508-9, 527	4/88—5/88	£737

The cylinders, motion and boiler remained unchanged but an open Stirling cab and separate rear bunker, similar to those used on the inside-cylinder 0–6–0STs, replaced the bunkerless Drummond cab of the 0–4–0STs. The tubes and inner firebox were steel while the buffers were the standard Drummond type. Originally, braking was entirely by hand. Dimensions peculiar to the class were:—

Wheelbase	13ft (7ft plus 6ft)
Length over buffers	27ft 10¼in
Water capacity	900 gallons
Coal capacity	2 tons
Weight in working order	31 tons ½cwt

The six engines were distributed between Polmadie, Edinburgh and Dundee. Although able to carry more coal and water than the 0–4–0STs, their longer wheelbase restricted them to a maximum curvature of 4½ chains. Since the much larger inside-cylinder 0–6–0STs could negotiate curves of the same radius, there was little to justify the construction of the less powerful engine. Not surprisingly, a return was made to the four-coupled type when additional dockyard shunters were required, two further batches of the standard 0–4–0ST being built by Drummond:—

Order No	Date ordered	Engine Nos	Dates delivered	Cost per engine
Y22	12/88	615-20	4/89–7/89	£667
Y27	5/90	510-15	4/90–5/90	£685

Nos 514-5 had steam brakes when new but the remainder were originally hand-braked.

The only passenger tank engines built during this period were a further six 171 class 0–4–4Ts:—

Order No	Date ordered	Engine Nos	Dates delivered	Cost per engine
Y19	8/88	222-7	2/89	£1230

These engines differed from the early batches in having the bunker extended to the rear of the platform. Water capacity was 855 gallons and coal 1¾ tons.

HIGH-PRESSURE TRIALS

A paper on compound locomotives read to the Institution of Civil Engineers in 1888 by Edgar Worthington provoked much discussion on the question of compound versus simple. Drummond pointed out that unless the engines under test had the same boiler and working pressure and used the same quality of coal, claims as to fuel economy were worthless. In reference to the work of No 123 during the Race to Edinburgh, he commented:—

> And its consumption in the running of this fast train, over gradients well known to be severe for a long part of the distance, never exceeded 31.6 lb of Scotch coal per train-mile; *and these were the same trains as are worked by the compound engines on the London and North-Western Railway.*

But it is with his final statement that we are most concerned:—

> To test the relative merits of compound engines as against ordinary modern engines, I shall be pleased indeed to ask permission of the Directors of the Caledonian Railway to be allowed to make arrangements to enable this to be done, either with the express passenger or goods engines of the Company, or both, the tests to extend over a period of not less than one month. This line, with its long and steep gradients between Carlisle and Aberdeen, is admirably suited for such a trial, and a test like this would settle definitely the question as to

IMPROVING THE BREED

which of the two descriptions of engine is the most economical in fuel and upkeep.

The Caledonian did not take up Drummond's offer. Instead, Drummond set out to determine whether the economies which were being claimed for compounding could be achieved in a simple engine working at a higher pressure than usual. This decision coincided with a need for further express engines for the Carlisle-Perth expresses and for the through engine-workings between Carlisle and Aberdeen begun in 1887.

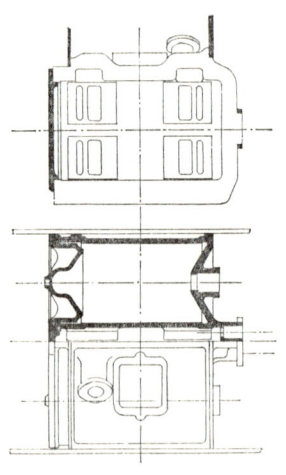

Fig 4 Cylinder arrangement : batch Y21

And so, in December 1888, six new express engines were ordered:—

Engine numbers	Order number	Dates delivered	Cost per engine
76-9, 84 & 87	Y21	7/89—11/89	£2107

Although outwardly similar to the 66 class, the engines embodied several ideas which put them far ahead of their time, the most important being the front-end arrangement. To secure short, direct steam passages, the steam ports were moved to the ends of the valve face. In place of the single exhaust port in the middle of the valve face, a separate exhaust port was placed beside each steam port, the total port area for exhaust thus being doubled. This

44

IMPROVING THE BREED

meant that a separate slide valve was required for each pair of ports, front and rear, the two valves being equivalent in function to the two heads of a piston-valve. The short, direct steam passages reduced the clearances and allowed rapid entry of steam into the cylinder, while the large exhaust port area permitted rapid exhaust. Against these advantages, the double slide valve weighed 70 per cent more than the ordinary valve and there must have been a considerable increase in frictional resistance.

As in the 66 class, all ports were divided horizontally into equal upper and lower portions, steam from the lower halves of the exhaust ports reaching the blastpipe via a passage running round the outside of the cylinder. In the new engines, the main frame acted as the outer wall of this passage, which was of greater cross-sectional area than that of the original 4-4-0s—35sq in against 21½sq in.

The slidebars and motion plate were arranged as in the Coast Bogies, the eccentric-rods being 11in longer than those of the 66 class. Reverse was by vertical screw of the type used on No 124; this enabled the driver to alter the cut-off by as little as 1 per cent. The steam circuit between the dome and the valve-chests was improved by increases in the diameter of the steam pipes and evaporation was assisted by the vortex blastpipe, which had an orifice equal in area to that of a plain pipe, 5in in diameter. The boiler was identical to that of the 66 class though, for the first time at St Rollox, mild steel was used for the plates of the barrel. Other departures from previous practice lay in the use of two No 11 injectors and in the ashpan, this having a single large damper underneath instead of the pair front and rear. Ancillary equipment included compressed air sanding and steam heating apparatus, both of the type used on the Coast Bogies. The tenders were of the new 3550-gallon pattern.

Four of the new engines, identical except for working pressure, were used in the trials. According to Drummond (*Proceedings of the Institution of Civil Engineers*, 1897), the engines selected worked at the following pressures:—

        Nos 76 & 79   200lb/sq in
        No 77   175lb/sq in
        No 78   150lb/sq in

These details do not accord with those given in the St Rollox boiler-proving register:—

IMPROVING THE BREED

Engine No	Working pressure	Date safety-valves set	Earliest date at which working pressure of 150 lb was recorded
76	175lb/sq in	11/7/89	5/95
77	200lb/sq in	25/9/89	10/95
78	200lb/sq in	9/10/89	6/95
79	200lb/sq in	26/10/89	11/91**
84	200lb/sq in	13/11/89	11/95
87	200lb/sq in	22/11/89	6/94

** To 160lb/sq in

It seems that the adjustments were made after 9 October and that these were not recorded in the register.

The trains selected for the trials were the 10.15 am express from Edinburgh to London and the corresponding down train, the 4.30 pm from Carlisle. The schedules demanded an average speed of 51.35mph in the down direction and 50.1mph on the up journey. Loads varied from 137 to 153 tons between Edinburgh and Strawfrank Junction and from 237 to 275 tons between Strawfrank Junction and Carlisle. A miners' strike was in progress and the only coal available was of very poor quality, half of it being small enough to pass through a $\frac{1}{2}$in mesh riddle; its evaporative power was only 6lb of water per lb of fuel fired.

The hardest work was done by No 79:—

Engine No	Edinburgh-Strawfrank	Strawfrank-Carlisle	Carlisle-Strawfrank	Strawfrank-Edinburgh
76	49.2	51.0	51.0	53.9
77	50.6	52.3	46.2	56.6
78	49.2	52.6	43.7	53.9
79	52.2	53.9	49.6	59.7

Average speed, mph

The maximum indicated horsepower developed was 940 (No 79) and the maximum speed attained was 81.8mph (No 76).

The engines with the higher working pressure worked on an earlier cut-off, the ranges being as follows:—

Engine No	Cut-off per cent	Engine No	Cut-off per cent
76	13-27	77	23-33
78	33-47	79	19-33

The results were assessed by comparing the amounts of steam used per indicated horsepower per hour, the outcome being a clear saving for the engines working at the higher pressure.

IMPROVING THE BREED

	Steam saved, per cent	
Engine No	Relative to No 77	Relative to No 78
76	11.9	31.0
77	—	15.0
79	10.2	24.5

The low steam consumption per ihp per hour—claimed to have been comparable to the best obtained with triple-expansion marine engines—and the fact that No 76 had been able to do some of her work on a five-fold expansion of steam provided Drummond with powerful arguments for the non-compound locomotive. 'As the whole question of engine economy resolves itself into the number of times steam can be expanded, and as in this case five expansions were within the economical limit in a single cylinder, compounding within this limit appears to be unecessary'. Much of his argument, however, was based on theoretical considerations and it was left to E. C. Poultney, in 1952 (*British Express Locomotive Development*), to underline the brilliance of the results. Details of the performance of Nos 76 and 78 are shown in Table 2, which

TABLE 2
PERFORMANCE: 1889 TRIALS

Engine No 76

Diagram No	1	5	6	9	10	12
Boiler pressure, lb/sq in	200	200	195	195	200	200
Regulator opening	¾	Full	Full	Full	Full	Full
Steam chest pressure, lb/sq in	200	200	195	195	200	200
Cut off per cent	21	18	18	27	18	13
Speed, mph	48	45	43	32	49	58
Mean effective pressure, lb/sq in	52.2	57.0	58.4	87.1	40.8	30.8
Indicated horse power	715	722	708	781	571	513
Steam, lb per IHP hour	18.4	17.7	17.8	19.1	17.5	17.3
Steam rate, lb per hour	13170	12760	12620	14880	9980	8810
Cylinder efficiency, per cent	11.5	12.0	11.9	10.9	12.2	12.3

Engine No 78

Diagram No	2	5	6	7	8	9
Boiler pressure, lb sq/in	150	155	150	150	150	145
Regulator opening	⅝	¾	⅝	Full	Full	Full
Steam chest pressure, lb/sq in	150	155	150	150	150	145
Cut off, per cent	33	33	33	33	33	47
Speed, mph	43	50	45	28	43	25
Mean effective pressure, lb/sq in	58.6	52.0	58.3	63.6	61.0	87.5
Indicated horse power	711	733	739	504	748	623
Steam, lb per IHP hour	23.5	24.2	22.3	22.5	22.4	23.5
Steam rate, lb per hour	16670	17680	16470	11300	16740	14550
Cylinder efficiency, per cent	9.0	8.8	9.5	9.4	9.5	9.0

includes the rates of evaporation as calculated by Poultney. When diagram 5 was taken, No 78 was using steam at a rate of 17,680lb per hour. Since full boiler pressure was being maintained, the high steam-rate must have been matched by an equally high rate of evaporation. This was a fine achievement for a small engine, particularly with poor coal. Fuel consumption varied from 42.3 to 53.2lb per mile, though had coal with the normal evaporative power of 8lb of water per lb fired been available, the consumption would probably have been around 35lb per mile.

In many ways, 1889 represented the high-water mark in Drummond's career. The new 4-4-0s embodied some of the most advanced thinking in British locomotive practice and their performance bore out the wisdom of those ideas. Nevertheless, Drummond could not persuade his drivers to make full use of the high pressure and refined front-end layout by working with full regulator and early cut-off:—

> Viewing the question of steam pressures broadly, I have come to the conclusion that for the present, until drivers appreciate the value and take advantage of higher pressures for ordinary locomotives working main line traffic economically in all respects, the pressure should not be less than 150 lb nor more than 170 lb per sq in.

And so when further 4-4-0s were required, a reversion was made to the front-end arrangement of 1884, only slightly modified. Indeed, Nos 77 and 79 were re-cylindered in August and November 1891 and it is likely that their new cylinders also were of this pattern.

Following the trials, the six engines settled down to regular work between Carlisle, Perth and Aberdeen. No 79 was named *Carbrook* after the estate of the chairman, J. C. Bolton, and was exhibited at the Edinburgh Exhibition of 1890. Under driver Tom Robinson of Carlisle, No 78 was to achieve renown for her performance in the Race to Aberdeen.

CHAPTER 4

THE TRANSITIONAL PHASE 1891-5

PERSONNEL CHANGES

The most significant event of 1890 was the resignation of **Dugald Drummond**. In May 1889, the chief commissioner of the New South Wales Railways had advertised for engineers to set up a works in New South Wales and had promised to order 75 locomotives every five years from any new company thus established. Along with three others, Drummond formed a consortium, the Australasian Locomotive Engine Works Ltd, which on 13 February 1890, offered to supply 100 engines within two and a half years. On 3 April, Drummond intimated his wish to resign from his post at St Rollox. He had been one of the most highly-prized locomotive engineers of his day; whereas the much-respected Benjamin Conner had received a salary of only £1100 per annum during his last years in office and Brittain a mere £850, Drummond had *begun* at £1700 per annum and had received an increase of £300 in August 1884 and a further rise of £400 in August 1888. The board's sense of loss was recorded on 15 April:—

> Resolved that Mr Drummond's resignation be accepted, the Board at the same time desiring to express their great regret that the Company should lose the services of an officer of whose value they are very sensible, attested as it is by the admirable state of efficiency of the Locomotive Works and of the Rolling Stock under his charge. Mr Drummond carries with him the warm regard of the Directors and their sincere wishes for his future happiness and prosperity.

He was also presented with a cheque for £1000 'in further ack-

THE TRANSITIONAL PHASE 1891-5

nowledgment of the Board's appreciation of his services to the Company'.

At the same board meeting, Hugh Smellie, the locomotive superintendent of the G & SWR, was appointed to succeed **Drummond** at a salary of £1500 per annum. Unfortunately, little time was available to him to express his undoubted engineering ability. Following a soaking received while attending to a derailment in Buchanan Street tunnel during the 1890-1 Scottish rail strike, he was taken ill and died on 19 April 1891.

His successor was John Lambie, appointed 2 June 1891, at a salary of £900 per annum. This heralded a change of policy on the part of the board. Drummond had been engineer first and foremost and his advanced and very sound locomotive designs had formed a sure foundation on which others might safely build. Lambie was a 'running' man; since September 1876 he had toured the system as assistant (outdoor) locomotive superintendent and had gained an intimate knowledge of CR engines and men. It is not surprising that he approached locomotive design with the needs of his enginemen in mind, his efforts on their behalf being exemplified by the introduction of cab-doors, longer handrails and additional footsteps. He achieved popularity with directorate, management and men and during his illness in 1894 received a warm message from Sir James Thompson. He died in harness on 1 February 1895. His final salary had been £1100 per annum but the board provided his widow with an allowance of £200 per annum under the trusteeship of William Patrick (later general manager) and J. F. McIntosh.

4-4-0 DESIGN AND PERFORMANCE

Drummond's last express engines were ordered in June 1890, the details being as follows:—

Order No	Engine Nos	Dates delivered	Cost per engine
Y25	83, 88-91 & 113	1/91—3/91	£2110

These engines differed from the Y21 batch mainly in their front-end arrangement but there were minor differences also, e.g., the crank-axles were made of steel and sanding was by gravity alone. Working pressure was originally 160lb/sq in but was reduced to 150lb/sq in in 1896-7. The front-end was based on that of the 1884

Page 51 (upper) Lambie 0–6–0 No 335 with Lambie 2840 gallon tender; *(lower)* 766 class 4–4–0 No 769 at Dundee

Page 52 (*upper*) Lambie 4–4–0T No 2 in original blue livery; (*lower*) 19 class 0–4–4T No 19 as repainted black

THE TRANSITIONAL PHASE 1891-5

engines, the only change being in the shape of the exhaust passage from the lower half of the exhaust port, the frame of the engine acting as its outer wall. The tenders were of the 3550-gallon pattern but lacked the hook-shaped spring-hangers.

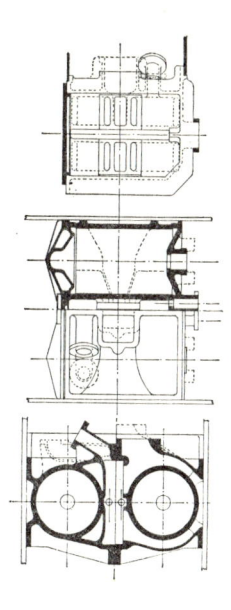

Fig 5 Arrangement of standard 18in × 26in cylinders : 1891

By 1893 the Drummond 4-4-0s had demonstrated their ability to master the heaviest loads which the traffic department could give them. No 124's *tour de force*, when the 73¼ miles from Carlisle to Strawfrank were run in 86½min with a load 'equal to fifteen', has already been mentioned. The best run in the up direction recorded by Pearson Pattinson* was made with a load of 'equal to nineteen', the details being as follows:—

	Miles	Time (min)	Average speed, mph
Carstairs	0	0	—
Beattock Summit	23¾	34	42
Beattock	33¾	43½	63
Lockerbie	47¼	57¼	61
Carlisle	73½	84¼	57

* *British Railways* by J. Pearson Pattinson. 1893.

C.R.—D

THE TRANSITIONAL PHASE 1891-5

Other performances recorded with heavy loads were:—

	Miles	Time (min) start-to-stop	Average speed mph	Load
Carlisle-Beattock	39¼	52	46	=23
Carstairs-Beattock Summit	23¾	38	37½	=23½
Carstairs-Carlisle	73½	92	48	=24

What speeds could be attained with a train of 'equal to nine' were shown when the 27½ miles from Edinburgh to Carstairs were run in 31min 17sec, start to stop, an average speed of 52.8mph. Up the long 1 in 100, speed never fell below 46mph and the summit at Cobbinshaw, 18¾ miles from the start, was passed in 22¾min. The average weight per vehicle on the best trains around 1890 seems to have been about 19 tons but the system of assessing loads, whereby an eight-wheeled coach was regarded as equal to one and a half six-wheelers, meant that a train of 'equal to twenty-four' could have weighed anything from 310 to 350 tons tare.

A further six 'Coast Bogies' were ordered in November 1890.

Order No	Engine Nos	Dates delivered	Cost per engine
Y28	114-15, 195-8	7/91—9/91	£1994

Drummond had resigned before these engines were ordered and Smellie made several important changes in the boiler design. A flat-topped inner firebox replaced the elliptical-roofed type used by Drummond, the crown being stayed as follows: Eight longitudinal girders were secured to the roof of the inner firebox, each being linked to two curved bars rivetted to the underside of the outer firebox casing. The dome was in the Drummond position and carried the safety-valves in the Drummond style but the tube arrangement was completely altered and the working pressure increased. The dimensional differences between the boilers fitted to the two batches of the class were as follows:—

Batch	Y13	Y28
Number of tubes	177	234
Diameter of tubes, in	1¾	1⅝
Heating surface: Tubes, sq ft	837.5	1028.6
Heating surface: Firebox, sq ft	99.0	111.0
Heating surface: Total sq ft	936.5	1139.6
Grate area, sq ft	16¾	17
Boiler pressure, lb/sq in	150	160

The boilers and crank-axles of the new engines were made of

THE TRANSITIONAL PHASE 1891-5

steel whereas those of batch Y13 were of iron. The cylinder design followed that of the Y25 batch of 6ft 6in 4–4–0s. Whether or not the vortex blastpipe was fitted is uncertain; a drawing dated February 1892 shows a plain blastpipe with an orifice of 4¾in diameter but the drawing office sometimes laboured under false assumptions and it is possible that the vortex blastpipe was again used. The tenders were of the Drummond 3130-gallon pattern. Several of these engines went to Perth and worked to both Crieff and Dundee. Edinburgh got two for the passenger turns to Carstairs, Lanark and Muirkirk. Nos 195 and 198 went to St Rollox shed, each being used for officers' specials during the 1902-5 period.

Six further express engines were ordered in May 1893.

Order No	Engine Nos	Dates delivered	Cost per engine
Y35	13-18	4/94—5/94	£1832

These were basically Drummond engines with Smellie boilers, Lambie boiler-mountings, additional handrails and footsteps and larger tenders. The cylinders and steamchests were identical to those of the 1891 engines though the steam pipes approached the valve-chests at right angles instead of at the more gradual angle employed by Drummond and the junction of the blastpipe with the exhaust passage round the outside of the cylinder was made more streamlined. The slidebars, motion-plate and eccentrics were arranged as in the later Drummond engines but reverse was by lever, the quadrant having eight notches on either side of mid-gear. The bogie was pivoted 1in in front of its centre-line and, as in the later Drummond engines, clearance for the rear bogie-wheels was provided by setting in the lower portion of each frame by 1in. Sanding was again by gravity alone.

The most noticeable changes were in the boiler-mountings. The safety-valves were placed on top of the firebox and the dome was brought further forward than before. But equally important were the internal changes derived from Smellie's brief occupation of the chair. As in the Y28 batch of 'Coast Bogies', already described, the shape of the inner firebox was altered, its flat roof being stayed to the outer firebox casing by longitudinal girders, and the tubing arrangement was changed, tubes of 1⅝in diameter being used instead of the 1¾in tubes favoured by Drummond. Lambie's own contributions included the Ramsbottom double-beat regulator and a new pattern of chimney with flat-topped rim. The vortex blastpipe was again employed, the equivalent diameter of its orifice being 4¾in. As in the earliest Drummond locomotives. dampers were fitted to the front and rear of the ashpan and a single organpipe whistle was used.

The appearance of the new engines was further modified by changes in various external fittings. The handrails alongside the boiler were made continuous with the one above the smokebox door, a small crank connecting the control rod inside the left-hand rail with the blower valve on the side of the smokebox. Footsteps were fitted to the running-plate opposite the rear bogie-wheels and handrails were attached to the engine and tender on either side of the cab entrance. The smokebox door was secured by a wheel and handle instead of the two handles previously used and the 'tallow cups' on the smokebox front were replaced by a pair of Furness lubricators which fed the cylinders when the engine was coasting. (A sight-feed lubricator in the left-hand corner of the cab supplied oil to the slide-valves.) The tender carried 3570 gallons of water and differed from the Drummond 3550-gallon pattern mainly in its shorter rear overhang and higher sides.

When new, one of the '13' class was tested on the Corridor from Glasgow to Carlisle and back. In the up direction the engine ran non-stop to Carlisle at an average speed of 45½mph with a load of 235 tons gross. The return journey was made with 212 tons gross, the highlight being the average speed of 54.2mph between Carlisle and the Beattock stop. Coal consumption for the round trip worked

TABLE 3
THE RACE TO ABERDEEN

Date	14/8	17/8	19/8	20/8	21/8	22/8	23/8
Carlisle-Perth							
Engine No	78	90**	90	78	90	78	90
Load, tons	172	207	188	95	95	95	72½
Time, minutes	167*	168*	165*	155*	155*	150*	149¼
Average speed, mph	54.3*	53.9*	54.9*	58.4*	58.4*	60.3*	60.4

* Inclusive of stop at Stirling
** Piloted by No 78 to Beattock Summit

Perth-Aberdeen							
Engine No		17		70	17	17	17
Load, tons		110		72½	72½	72½	72½
Time, minutes		96¼*		89	83	84	80½
Average speed, mph		56.1*		60.0	64.8	64.1	66.8

* Inclusive of stop at Forfar

THE TRANSITIONAL PHASE 1891-5

out at 35.02lb per mile and water consumption at 26.3 gallons per mile, 7.51lb of water being evaporated per lb of coal fired.

The Race to Aberdeen has been the subject of a special study by O. S. Nock and further description would merely be repetitive. A summary of some of the most noteworthy performances, including those put up during the final four nights, is given in Table 3.

Rous-Marten was impressed by the performance of the CR engines, his 'East Coast' leanings during the race giving way to a deep and lasting enthusiasm for the products of St Rollox. In *The Engineer* of 28 February 1896, he wrote:—

> But superb as were their performances during that memorable period, feats of equal merit were almost daily achieved with the ordinary expresses. In illustration of this, I mention the following:— No 87 took the West Coast corridor dining train—equal to twenty-two coaches—from Symington to Carlisle, just under 67 miles, in 72min 57sec. The 17 miles to Beattock Summit . . . occupied but 22min 13sec, and the speed did not fall below 35 miles per hour on the last two miles of 1 in 100 up. The 50 miles from Beattock Summit to Carlisle were done in 50¾min. No 14 took the corridor train—equal to seventeen coaches—from Symington to Carlisle in 71½min; No 79 took seventeen coaches from Beattock Summit to Strawfrank, 23¼ miles—start to stop—in 23min 11sec; No 75 took seventeen coaches from Forfar to Perth, 32½ miles, in 32min 59sec; No 67 made the same run with thirteen coaches in 32min 41sec, and with ten coaches in 32min 24sec, all being from start to stop. Such work as this speaks for itself.

The complete allocation of the Drummond and Lambie 4-4-0s during the 1890s cannot be established with certainty. During the late 1880s, Nos 68-9 were at Stranraer and Nos 73-4 at Perth but the length of their stay at these sheds is unknown. In 1895 the known allocations were:—

Aberdeen	Nos 13-15
Perth	Nos 16-18, 61, 66-7, 70-2, 75-6
Polmadie	Nos 63-5
Ardrossan	No 124
Edinburgh	Nos 60, 62, 77, 113
Carlisle	Nos 78-9, 89-91

0-6-0 DESIGN AND CONSTRUCTION

Although the 1890s are best known as the 'decade of the Dunalastair', an equally important development at St Rollox was the construction of large numbers of six-coupled goods engines, the 'Jumbos' reaching a final total of 244 in December 1897. As the

THE TRANSITIONAL PHASE 1891-5

class expanded, modifications were made to the design of boilers, tenders, etc., though the actual dates at which the various changes occurred are sometimes difficult to establish. Drummond's last 0-6-0s were built to order No Y23 as follows:—

Date ordered	Engine Nos	Dates delivered	Cost per engine
6/88	403-15	11/89—7/90	£1756

These engines were the last of the class to have iron crank-axles. Although a new type of tender was designed for them (see Chapter 3), and fitted to No 404 at least, all eventually received second-hand tenders, No 413 acquiring the original Neilson tender from No 123 the remainder being allocated Stroudley-type tenders displaced from the 4-4-0s. Nos 410-15 were fitted with the Westinghouse brake and the remainder with steam brakes.

The next three batches of the class were constructed to Smellie's specifications.

Order No	Date ordered	Engine Nos	Dates delivered	Cost per engine
Y29	2/91	372-9, 540-3	11/91—2/92	£1805
Y30	10/91	691-6	4/92—5/92	£1888
Y31	12/91	544-53	7/92—8/92	£1805

The boilers fitted to Nos 372-3 were built to order M70.

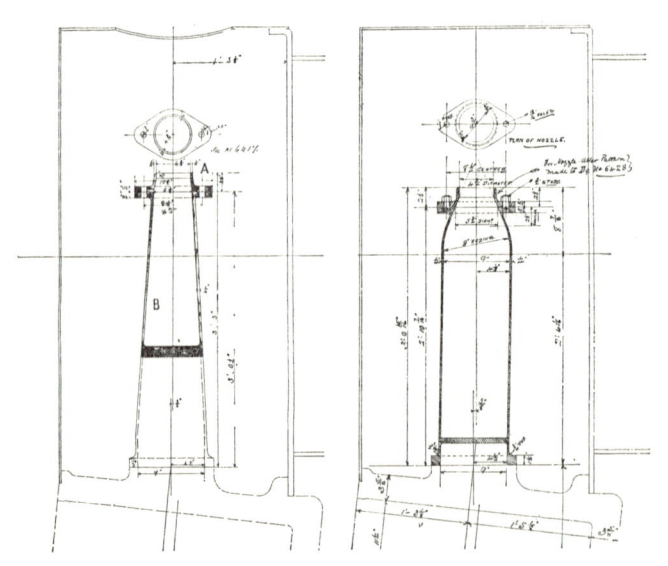

Fig 6 Blastpipe variation: *Left* batch Y29 *Right* batch Y30

THE TRANSITIONAL PHASE 1891-5

All of these engines differed from their Drummond predecessors in having flat-topped inner fireboxes with longitudinal girder-stays and 242 (later 238) tubes of $1\frac{5}{8}$in diameter. A new pattern of blast-pipe, $2\frac{1}{4}$in lower than the Drummond type but with the same $4\frac{3}{4}$in diameter orifice, was designed for the Y30 and succeeding batches. Nos 691-6 were Westinghouse fitted but the remainder had steam brakes only and were painted black. All three batches received Stroudley-type tenders of 2500 gallons capacity.

Next followed two batches with Lambie modifications. These were:—

Order No	Date ordered	Engine Nos	Dates delivered	Cost per engine
Y32	5/92	554-63	12/92—1/93	£1745
Y33	6/92	697-702	3/93—5/93	£1879

The cylinders fitted to these and to all succeeding batches of the class were of modified design, the arrangement of the steam-pipes and exhaust passages being the same as in the 13 class 4-4-0s. The diagram book and official drawings show these engines to have had Lambie boilers with the safety-valves on the firebox and the dome set further forward than before. The cab and boiler hand-rail arrangements, the blower and the smokebox door fastening were the same as on the 13 class. For greater safety, a handrail-plate was fitted to each side of the tender footplate and a door, hingeing on to each plate, completely covered the gap between cab and tender. The latter carried 2500 gallons of water but whether or not it was of the 1889 (Y23) type is uncertain. Batch Y32 were equipped with steam brakes but the remainder were Westinghouse fitted. No 698, which went to Aberdeen, was later vacuum-fitted for working 'foreign' fish stock.

The rest of the Lambie engines were built as follows:—

Order No	Date ordered	Engine Nos	Dates delivered	Cost per engine
Y36	5/93	199-202	2/94	£1879
Y37	12/93	256-61 334-7, 203-4	5/94—11/94	£1638
Y38	1/94	703-10	10/94—6/95	£1638

These engines had the same boiler, with 238 $1\frac{5}{8}$in tubes, as the earlier Lambie engines but the dome covers fitted to batches Y37 and Y38 were slightly larger than those used previously. In style as well as in detail, these engines were pure Lambie. The tenders

THE TRANSITIONAL PHASE 1891-5

had deep frames with springs above the axleboxes and the tank was deeper than before but of the same length, with rear toolbox. The well was shorter than in the old 2840-gallon tenders but the deeper tank enabled the water capacity to be maintained at 2800 gallons. The handrail-plates fitted to the tenders of batch Y36 were identical to those designed for batch Y32, the top being formed of a straight angle-iron. From batch Y37 onwards the rear of the angle-iron mirrored the cab 'lean-out' by curving upwards to meet the tender side and the handrail was repositioned slightly. Nos 199-202 were Westinghouse-fitted but the remainder had steam brakes and were painted in the goods black livery.

TANK ENGINES: 1891-5

In 1891 a final batch of six 171 class 0-4-4Ts was built at St Rollox.

Order No	Date ordered	Engine Nos	Dates delivered	Cost per engine
Y26	5/90	189-194	4/91-5/91	£1265

These engines had full-length bunkers, vortex blastpipes, twin whistles and train-heating equipment arranged on the Drummond plan. The steam from the Westinghouse pump exhausted into the train-heating pipe whereas that of the earlier engines had been piped alongside the boiler before being discharged into the smokebox. The class became closely identified with the rural branch lines and eventually could be seen from the Solway Junction Railway in the south to Forfar and Arbroath in the north. Later, many of the class became station pilots at such places as Perth, Forfar, Dundee and Edinburgh.

In 1895, five inside-cylinder 0-6-0STs were built to Lambie's design.

Order No	Date ordered	Engine Nos	Dates delivered	Cost per engine
Y39	6/94	211-15	1/95-3/95	£1253

These engines were based on the earlier saddle-tanks but had fully enclosed cabs of Drummond type. The 'rebuild' boiler was again used, but with $1\frac{5}{8}$in tubes and Lambie boiler mountings. The dome was brought further forward than before and the safety-valves and whistle were transferred to the firebox, this necessitating a shortening of the saddle-tank which now terminated well ahead of the cab. The chimney, smokebox, handrail, blower and lubricator

THE TRANSITIONAL PHASE 1891-5

details were of standard Lambie type but cast-iron wheels and Conner buffers were fitted. Water capacity was 950 gallons and coal 2 tons. The engines were equipped with the steam brake, painted black and allocated to Grangemouth for dockyard duties. As a much-improved version of the Drummond engines, they soon acquired the nickname 'New Jubilees'.

During the middle 1890s, St Rollox was faced with providing suitable engines for two new lines, the Lanarkshire and Dumbartonshire Railway and the Glasgow Central Railway, then in process of construction. The major feature of the new system was the long underground section between Dalmarnock and Stobcross. Work on this began in June 1890 but it was not until August 1896 that the Glasgow Central Railway was opened throughout; the Lanarkshire and Dunbartonshire, which shared the underground tracks from Glasgow Central to Stobcross, following on 1 October.

Several classes of locomotive were specially built for the 'low-level' lines, each being fitted with condensing apparatus to minimise pollution. The earliest was Lambie's 1 class 4-4-0T, turned out as follows:—

Order No	Date ordered	Engine Nos	Dates delivered	Cost per engine
Y34	12/92	1-12	7/93−1/94	£1614

The frames, bogies and springs were arranged as in the 4-4-0 tender engines though bogie wheelbase was reduced by 6in and bogie-wheel diameter by 4in. A new front-end, with 17in x 24in cylinders and undivided ports, was introduced but the diameter of the coupled wheels remained at 5ft as in the Drummond 0-4-4Ts. The slidebars were separate from the cylinders and were supported by the motion-plate in the mid-length position. The boiler was of the standard Lambie 'rebuild' type and was fed by a Drummond injector in the usual position on the right-hand side and by a crosshead pump on each side, the right-hand clack-valve being double. The injector steam cock was of the type used on the Drummond tank engines.

Nine of the class had gravity sanding front and rear, the forward gear being operated by a vertical lever and quadrant on the left-hand side of the footplate and the backward gear from a horizontal sector on top of the sandbox in the left rear corner of the cab. The remaining three, including No 10, were fitted with Gresham and Craven steam sanding equipment, a two-way cock below the injec-

THE TRANSITIONAL PHASE 1891-5

tor steam cock allowing the driver to apply sand to the leading or to the trailing coupled wheels. The driver's brake valve was attached to the faceplate as in the 4-4-0 tender engines. The brake cylinder and auxiliary Westinghouse reservoir lay below the footplate and applied the brake-blocks to the front of the wheels by outside brake-rods. There were two main reservoirs, located beneath the running-plate at the front of the engine. Reverse was by lever and fore-end lubrication by a sight-feed lubricator in the front left-hand corner of the cab and by a pair of Furness lubricators on the smokebox front. The bunker, which lacked the usual shallow tank, was wider than the cab, its corners being rounded and its sides flared out at the top. Water capacity was 1000 gallons and coal 2 tons.

The blastpipe was conventional but at its base was situated an intercepting valve which diverted exhaust steam to the bottom of the tanks via an oblong box on top of each running-plate. The exhaust was then piped to an inverted U-bend on top of the tank before being discharged further down. Any steam which remained was allowed to escape through a standpipe in front of the cab. Initially, the intercepting valve was operated by a wheel and screw at the rear of the right-hand tank, the control-rod running through the main handrail to a crank on the side of the smokebox, but the screw was soon replaced by a rack and pinion operated by a lever. In 1897 one engine was fitted with a 'pet cock' controlled from the cab by a brass sector and lever, this allowing the return of water from the pumps to the tanks when the boiler was full.

The engines were painted blue. Two were sent to Airdrie and the remainder, it is thought, to Polmadie. When the 'low-level' lines were opened, the latter group were transferred to Dawsholm. For some time thereafter, their main employment was on the Maryhill-Airdrie services, the Airdrie engines sharing these workings as well as operating the Airdrie-Newhouse branch. By 1898 all twelve engines had been repainted in the goods black livery which was more suited to the underground conditions.

Operation of the condensing apparatus in the underground section was strictly enforced but since the steam pressure generally fell and the injector, and sometimes even the pumps, refused to deal with the hot feed, many drivers 'forgot' to bring the condenser into use and, as a consequence, had to face disciplinary action. At one stage, 15 drivers were reported for this offence within three

THE TRANSITIONAL PHASE 1891-5

weeks and six habitual offenders were each sentenced to one day's suspension. When the condenser was used, hot water had frequently to be discharged from the tanks and replaced with cold water before the boiler could be filled. Several protest meetings were held by the drivers concerned, it being stated that 'due to the continual running, any stopping time was taken up by filling the tanks with cold water and there was no time to eat food on the job'.

The next class built for the 'low-level' lines was Lambie's last design. In this, reversion was made to the 0-4-4 type, ten engines being constructed as follows:—

Order No	Date ordered	Engine Nos	Dates delivered	Cost per engine
Y40	6/94	19-28	7/95—8/95	£1700

The boiler was the same as that used on the 4-4-0Ts but the diameter of the driving wheels was increased to 5ft 9in and standard 18in x 26in cylinders with divided ports were fitted. The smokebox was lengthened by 5½in and the steam circuit improved by increases in the diameter of the steam-pipes. The slidebars were attached to the cylinders in front and were supported by the motion-plate at the rear, this necessitating an 11in reduction in the length of the eccentric-rods. Laminated springs were used for the leading axle and spiral springs for the crank-axle. The side-sheets of the Lambie-style bunker were lengthened by 2ft 2in but, since over half the available volume was occupied by a water-tank, coal space was rather restricted. Sanding was by gravity, a single lever in a quadrant on top of the sandbox in the left rear corner of the cab supplying sand for forward as well as backward running. A second lever in the same quadrant opened the cylinder-cocks. The Westinghouse brake-valve was again attached to the faceplate but the main reservoirs were below the running-plate at the rear of the engine. Reverse was by lever, with six notches on either side of mid-gear. Water capacity was 1000 gallons and coal 2 tons.

The intercepting valve diverted exhaust steam to two copper pipes which ran back alongside the boiler to a small dome halfway along each tank. From there steam passed down a perforated pipe to the bottom of the tank, the pipe continuing backwards below the running-plate to the bottom of the bunker-tank. Any uncondensed steam in the side-tanks escaped through two standpipes in front of the cab windows. The intercepting valve was

THE TRANSITIONAL PHASE 1891-5

operated by the same rack and pinion as fitted to the 4-4-0Ts but the control-rod passed alongside the boiler clothing instead of within the handrail. While the new engines were decidedly superior to the 4-4-0Ts, they did share some of the latter's faults. Crosshead pumps, differing only in stroke from those fitted to the 1 class, and a Drummond injector on the right-hand side again supplied the boiler, the clack-valves, injector steam cock and watercocks being identical in layout and design to those of the earlier engines.

The engines were painted blue and stored at St Rollox until the line was opened. According to McIntosh, they were intended 'to run the express trains to Dumbarton and Balloch through the underground railway from Glasgow stations'. After a short spell on these duties they were repainted in the goods black livery and later joined the 4-4-0Ts on the more prosaic duties.

RENEWALS

Some boilers ran for long periods before repairs were required; that of No 316, for example, ran for nine years before major repairs were necessary, the only work carried out during that time being the replacement of nine tubes. The 4-4-0s ran much higher mileages than the other classes and therefore required more frequent renewal. The following boiler repairs are representative of the two extremes of wear.

Engine No	Date	Renewals
60	6/88	Half copper tubeplate. 226 new brass tubes.
	4/89	226 new brass tubes.
	12/89	Half copper tubeplate. 226 new brass tubes.
	2/92	Half copper tubeplate. 226 new brass tubes. New copper firebox crown.
	5/94	Replacement boiler (see below).
70	4/90	216 new brass tubes.
	6/91	14 new brass tubes.
	9/92	Half copper tubeplate. 226 new brass tubes.
	10/94	Replacement boiler (see below).

A noteworthy development in 1893 was the establishment of a system of boiler interchange, designed to curtail the time engines spent in the works during major overhaul. This was facilitated by the construction of a number of spare boilers in 1895 (order No Y41). The interchanges occurring up to February 1897, beyond which records are not available, were as follows:—

64

THE TRANSITIONAL PHASE 1891-5

Old boiler from engine No	Fitted to engine No	Date
72	75	1/93
75	68	6/93
68	71	10/93
71	74	2/94
74	60	5/94
60	70	10/94
70	66	7/95
66	316	10/95
316	311	12/95
New boiler	367	10/95
367	362	12/95
362	304	2/96
New boiler	61	12/95
61	60	4/96
60 (second boiler)	67	2/97
519	75	5/96
524	73	10/96
68 (second boiler)	307	10/96

The only engines known to have been re-cylindered prior to March 1897 were No 300 (12/91), No 305 (8/91), No 72 (9/91) and 77 and 79 as already recorded. In 1891 Nos 60-2, 65-7, 69, 71-6, 81-2, 123-4, 312, 321-2, 345-6, 360 and 368 received new crank axles. During the middle to late 1890s, the Drummond 4–4–0s were brought into line with their successors by the fitting of Furness lubricators, front footsteps, continuous handrails and the Lambie form of smokebox-door fastening.

During the early 1890s, further modifications were made to the design of tender axleboxes. A new method for lubricating the lower surface of the journal was introduced, an oil-pad being held against the journal by a coil-spring immersed in an oil-pan at the bottom of the axlebox, trimmings being used to convey oil from the pan to the pad. Various methods were used to lubricate the crown of the axlebox bearing and some of these applied to middle axleboxes only. In 1893, No 79's tender was fitted with concave journals in place of the cylindrical type previously employed. In the '13' class, concave journals were used on the leading and trailing axles only, but from order No Y47 until the end of the McIntosh regime they were applied to all three tender axles. Oscillation of the tender was reduced and the tendency of cylindrical journals to heat at the collars was obviated.

The 1890s also saw the introduction of several new pieces of equipment. In 1892 all six 4–4–0s of batch Y25 and 'Jumbo' No 411 were fitted with vacuum ejectors and piping, the two brake systems being synchronised by a rod which connected the handle of the Westinghouse brake valve to the vacuum brake valve. The

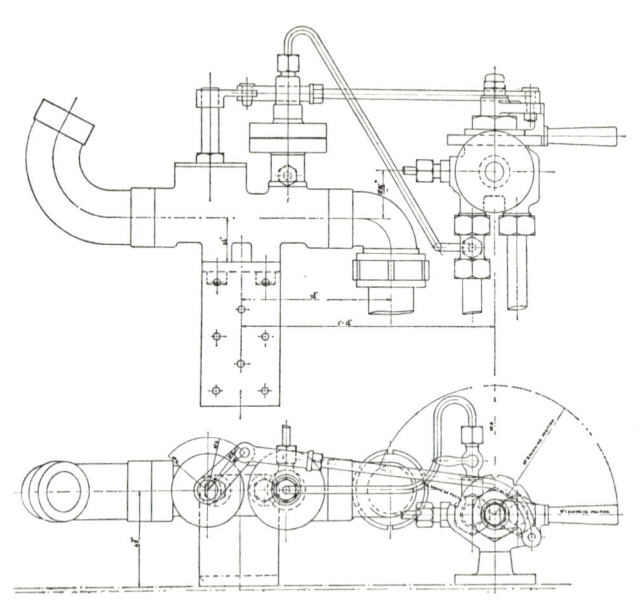

Fig 7 Connection between Westinghouse and vacuum driver's brake valves

latter, which had no handle of its own, consisted of the usual rotary, disc-shaped rose valve admitting air to the train pipe through a number of small holes (the rose). When the driver wished to apply the Westinghouse brake to the engine and the vacuum brake to a train of vacuum-fitted vehicles, he put his Westinghouse brake handle into the 'brakes on' position. Because of the linkage with the vacuum brake valve, this action allowed air to enter the vacuum train pipe and apply the vacuum brake. The two brake systems were equalised by an automatic valve (see Fig. 7). When the Westinghouse brake handle was moved to the charging ('brakes off') position, the rose was closed and the small

THE TRANSITIONAL PHASE 1891-5

ejector, which was kept continually in operation, recreated the vacuum. This apparatus remained standard for dual-fitted engines until 1917.

In November 1894 further thought was given to the subject of train-heating, and drawings were prepared showing apparatus on the 'Consolidated Carriage Company's System' applied to a Lambie 4-4-0. This system differed from the Drummond one in using live steam only, this being taken from the bottom of the Westinghouse steam cock and piped to a pressure-reducing valve on the faceplate before entering the train-heating pipe. It was not until 1897, however, that train-heating equipment was applied as a routine to all new express engines.

CHAPTER 5

THE RISE OF J. F. McINTOSH

THE NEW STAR

Lambie was succeeded by a man of very similar background. In February 1862, as a lad of seventeen, John Farquharson McIntosh had begun his apprenticeship in the Arbroath workshops of the Scottish North Eastern Railway. Promotion came rapidly and in 1876 he was made a locomotive inspector of the northern area. He soon rose to the rank of district locomotive foreman, first at Aberdeen, then at Carstairs and finally, in July 1886, at Polmadie, where his salary was £208 per annum. He succeeded Lambie as assistant locomotive superintendent in April 1891 and as locomotive superintendent in February 1895:—

> After full consideration, unanimously resolved to appoint Mr John F. McIntosh as successor to Mr Lambie at a salary of £700 P.A. The appointment to be tentative and for reconsideration at the end of 6 months.

The decision was made by the locomotive and stores committee, and not by the full board. This, together with the tentative nature of the appointment and the low salary which it commanded, suggests that the board wanted continuity, rather than change, at St Rollox. In explaining the appointment to the staff there, Sir James King said:—

> Instead of going outside for a man with great fame and perhaps a reputation which would not be borne out by experience when he came to the works, we took one of ourselves, and I am sure we have no reason to regret the choice.

McIntosh's experience had made him an intensely practical

Page 69 (*upper*) 'Coast Bogie' No 80 with 2840 gallon tender of late Drummond design; (*lower*) Dummond 4–4–0 No 70 as rebuilt with large boiler. Drummond 3550 gallon tender

Page 70 (*upper*) 439 class 0-4-4T No 222 outside Buchanan Street; (*lower*) 782 class 0-6-0T No 245 at Carstairs

THE RISE OF J. F. MCINTOSH

'running' man and he knew exactly how to develop the Drummond locomotive to make it the ideal 'engineman's engine'. His first express engines put a 60mph booking into *Bradshaw* for the first time ever and his sustained and rapid progress during the years 1896-1902 and his flair for publicity won him the admiration of the railway press. By 1900 he was earning £1500 per annum.

McIntosh's former position as chief outdoor assistant was filled by Tom MacDonald, previously district lomomotive foreman at Polmadie. The venerable Joseph Goodfellow retired in January 1895 and was succeeded by Peter Drummond who, at a salary of £450 per annum, remained in charge of the works until October 1896. Strangely enough, Lambie and Goodfellow lie in adjacent graves at Sighthill, a stone's throw from St Rollox. When Peter Drummond left for Inverness, he was succeded by R. W. Urie, the post of chief draughtsman then passing to William Urie.

THE LAST 'JUMBOS'

When McIntosh took command, the last batch of 0–6–0s (order No Y38) ordered by Lambie had still to be completed. Before the last two engines were finished, the decision was taken to fit them with the Westinghouse brake. It has been suggested that this was because almost 160 goods engines had had to be temporarily laid-off during the Scottish miners' strike of 26 June to 15 October 1894; without continuous brakes, these engines could not be used for passenger work. And so Nos 709-10 were Westinghouse-fitted and painted blue, as were a further 81 engines ordered by McIntosh.

Order No	Date ordered	Engine Nos	Dates delivered	Cost per engine
Y41 sup	12/94	711-20	6/95–8/95	£1714
Y45	10/95	736-60	9/96–4/97	£1837
Y46	11/95	564-75	5/96–7/96	£1714
Y47	4/96	576-82	11/96–12/96	£1837
Y47	4/96	583-7	2/97	£2001
Y49		761-5, 588-92	4/97–9/97	£1896
Y50		329-33, 593-9	9/97–11/97	£1896

Externally, these engines differed from their immediate predecessors in having inside brake-rods and small footsteps at the rear of the tender frames. Internally, the main changes were in the tubes (218 of 1¾in dia) and in the motion-plate arrangement. The slide-bars were attached to the cylinders and were supported at the rear by the motion-plate, this necessitating a reduction of 11in in the length of the eccentric-rods. Batches Y41 and Y45 had the vortex blastpipe but a return to the pattern introduced with batch Y30 seems to have been made for the later engines of the class. Gresham and Craven No 9 combination injectors were fitted to Nos 750-3 and may also have been used on some of the later engines for they are shown in a general arrangement drawing of November 1897.

Nos 583-7 were built for working over the 'low level' lines of the Glasgow Central Railway and were fitted with condensing apparatus. Exhaust steam was diverted from the blastpipe by an intercepting valve controlled from the cab and was carried back along each side of the boiler in a large polished copper pipe which, on entering the cab, dropped below the footplate to join a single pipe running along the bottom of the tender, this conveying the exhaust to a small dome on the rear of the tank. The tank-filler was given an air-tight lid and, for the first time, a water-level indicator was mounted on the front of the tender. Boiler-feed was by a Westinghouse feed-pump on the left side of the firebox and a Drummond No 11 injector in the usual position on the right-hand side.

TANK ENGINES: 1895-1900

McIntosh's first tank engines were four 0-4-0STs, very similar to the Drummond pugs but having cast-steel wheels with twelve spokes of normal design.

Order No	Date ordered	Engine Nos	Dates delivered	Cost per engine
Y43	4/95	611-4	10/95—11/95	£803

In these, steam brakes were fitted from the start.

Following closely on the heels of the saddle-tanks came nine 0-6-0Ts for the 'low-level' lines.

Order No	Date ordered	Engine Nos	Dates delivered	Cost per engine
Y42	4/95	29, 203-10	11/95—1/96	£1502

The wheelbase was the same as that of the Lambie 0-6-0Ts but the rear overhang was shortened by 6in and the wheels were of a new pattern with cast-steel centres and crescent-shaped balance-weights. The cylinders were the same as those of the 19 class but the 'rebuild' boiler was of McIntosh design, with 206 1¾in tubes.

THE RISE OF J. F. MCINTOSH

The side-tanks were much higher than those of the 19 class and the bunker was of the Drummond type, flush with the cab. The boiler was fed by two crosshead pumps and a single Drummond injector, the right-hand clack-valve being double. The intercepting-valve was similar to that of the 19 class but the control-rod ran through the handrail on the side of the boiler. The exhaust pipes from the smokebox ran straight back to the top of the side-tanks, the remainder of the piping being arranged as in the 1 class. Water capacity was 1300 gallons and coal 2½ tons.

The engines were Westinghouse-fitted and painted in the goods black livery. They entered service before the 19 class and were employed on goods and mineral trains on the completed sections. They were also used on the workmen's trains which served the shipyards along the north bank of the Clyde.

The last engines built for the underground lines were the 22 0-4-4Ts of the 92 and 879 classes, these being turned out as follows:—

Order No	Engine Nos	Dates delivered	Cost per engine
Y48	92-103	5/97—7/97	£1867
Y59	879-86	1/00—2/00	£2177
Y60	437-8	3/00	£2176

These engines were based on the 19 class, the main differences being:

(1) The side-tanks were of the same height as those of the 29 class, the condensing pipes being arranged as in the latter class. Water capacity was 1170 gallons.
(2) The bunker was of the Drummond pattern. (Coal capacity 2¼ tons.)
(3) The control rod to the intercepting valve ran through the handrail on the side of the boiler.
(4) The numberplates were of a new pattern with raised letters and numerals.

In batch Y48 the crosshead pumps were replaced by a single Westinghouse feed-pump, mounted on the running-plate, directly opposite the brake pump. This fed the left-hand clack-valve, the right-hand valve taking water from a Drummond No 11 injector. The boilers of this batch were of Lambie pattern with 224 1⅝in tubes. The later batches were regarded as a separate class, a fresh diagram being issued for them. They had McIntosh boilers with

206 1¾in tubes, crosshead pumps and front footsteps with one, instead of two, steps.

The 92 and 879 classes were undoubtedly the best of the condensing tank engines and were rather more widely distributed than the earlier engines, Hamilton and Motherwell each receiving an allocation for the operation of through trains to the underground lines from places as far away as Bothwell, Hamilton and Strathaven. All were painted black but several, including Nos 94, 97 and 102-3, were lined out with a white line on either side of a red band, as in the goods engine boiler bands.

DUNALASTAIR

The penultimate step in the transition from Drummond to McIntosh was made in July 1895 when 15 more express engines were ordered. Due to the introduction of corridor coaches and dining cars from 1893 onwards, the best Anglo-Scottish trains weighed more than in 1890 and, following the Race to Aberdeen, this increase was matched by a sharp rise in booked speeds. While McIntosh and his staff were working out new designs to meet these demands, the immediate need for more powerful engines was met by augmenting the boiler power of the 15 engines on order. As much of the remainder of the design was identical to that of the Lambie engines, only a limited number of new drawings were required and the general arrangement was not prepared until March 1898, some two years after the engines took the road. Interim measure though their construction was, the 15 engines soon established themselves as one of the most outstanding designs ever to have run on British metals. The first of the class, No 721, was named *Dunalastair* after the estate of J. C. Bunten, then deputy chairman, and the name was soon on the lips of almost every railway enthusiast in the land.

The details of construction were as follows: —

Order No	Engine Nos	Dates delivered	Cost per engine
Y44	721-35	1/96—5/96	£2104

The front-end arrangement, including the steam and exhaust connections, was the same as that used on the Lambie engines but the diameter of the cylinders was increased to 18¼in, this enlargement necessitating a reduction in the width of the exhaust passage from the lower portion of the exhaust port. The slide-bars were

THE RISE OF J. F. MCINTOSH

attached to the cylinders and were supported at the rear by the motion-plate, this necessitating a reduction to 4ft 7in in the length of the eccentric-rods, with corresponding effects on the valve-events. Reverse was by lever, the quadrant having eight notches on either side of mid-gear. As in the Lambie engines, the bogie pivot was 1in ahead of the bogie-centre and additional clearance for the rear bogie-wheels was provided by 'setting in' the lower part of the frames over the appropriate distance. The driving-wheels were of a new design with crescent-shaped balance-weights in place of the square-ended type previously used. The brake arrangement was completely altered, the brake-blocks being applied to the front of each driving-wheel by rods operated by a single cylinder slung vertically below the footplate. This required transfer of the main Westinghouse reservoir from the engine to the tender. Gravity sanding was again employed.

As with the 13 class, the main changes lay with the boiler. The barrel was pitched 6in higher than before and its diameter was increased by 3in, these changes creating an impression of bulk and power which contrasted with the slender grace of the earlier engines. The large boiler allowed an increase in the number and diameter of the tubes (265 of $1\frac{3}{4}$in), heating surface rising by 20 per cent. The firebox was of the type used on the 13 class but was $4\frac{1}{2}$in deeper at the front and $2\frac{5}{8}$in longer. The height of the boiler necessitated some modification of the exhaust arrangements, the vortex blastpipe (equivalent diameter $4\frac{13}{16}$in) being redesigned and a new type of chimney introduced. The latter was double-walled, the inner wall being formed by a liner which extended down into the smokebox to form a bell-shaped structure 7in deep and known at St Rollox as a 'hood'. (A conical hood had been tried experimentally on a 'coast bogie' in 1895.) The boiler mountings were arranged as in the Lambie engines, the dome housing a Ramsbottom double-beat regulator. Innovations included an index and pointer for the regulator handle*, Gresham and Craven No 9 combination injectors on the faceplate and McIntosh's patent water-level gauge protector which consisted of a cylinder of specially toughened glass, capable of withstanding a pressure of 3000lb/sq in, enclosing the gauge. The cab was an improved version of the Drummond type, the 'lean out' being 5in shorter. The combination injectors simplified the appearance of the engine

* First engine only

THE RISE OF J. F. McINTOSH

and the clean lines enhanced the bold proportions of the boiler, cab and Lambie 3570-gallon tender. However, in such external details as the smokebox door fastening, Furness lubricators, handrails and footsteps, the engines were much in the Lambie style.

The Dunalastairs were originally allocated as follows: Aberdeen No 726; Perth Nos 722-5; Polmadie Nos 721, 734-5; Edinburgh No 731; Carlisle Nos 727-30.

Dunalastair was run in by driver J. Ranochan on the slower or lighter trains between Glasgow and Carlisle. During this time, Rous-Marten was privileged to have a trip on her footplate and afterwards described the ease with which she handled her thirteen coaches, about 200 tons gross.

> This steep bank (Beattock) was ascended under easy steam, the regulator being only one-third open and the reversing lever in the fourth notch. . . . The lowest speed up the 1 in 75 was 25.7 mph but the engine was simply "playing with" the train, and working a long way within her powers . . . a short spurt was made down the length of easier falling gradient toward Lamington. The last three miles before this point were done at 73.8, 76.3 and 79.0 miles an hour respectively.

Running-in successfully completed, *Dunalastair* was allocated to the 2 pm Corridor from Glasgow Central to Carlisle and back. No 731 was put on the 10.15am from Princes Street to Carlisle, returning with the 4.30pm. This roster demanded an average speed of 52½mph in the up direction and 50.3mph in the down. The Carlisle engines ran the principal Glasgow trains other than the Corridor and also alternated with Perth on the very strenuous down Tourist, leaving the border city at 1.54am. This was booked to stop at Stirling in 125min and to cover the Stirling-Perth section in 35min, average speeds of 56.5 and 56.7mph respectively. The engine returned home with the 8pm sleeper from Perth. The Perth engines had four rosters:—

(1) The 8pm from Perth to Carlisle and the down Tourist from Carlisle. When this duty was worked by a Carlisle engine, the day's work consisted of an early morning and an early evening local to Dundee.

(2) The 9.10am Perth to Glasgow express.
A Glasgow-Cumbernauld local and back.
The 5.30pm express from Glasgow to Perth.

(3) The 12.6pm Perth to Glasgow.
The 5pm Glasgow to Dundee.
Dundee-Perth local.

(4) The 4.4pm Limited Mail, Perth to Carstairs.
The 10.24pm Carstairs to Perth, the Aberdeen portion of the down Corridor.

Two engines and two sets of men alternated on rosters 1 and 2. No 726 is thought to have run the 5.40pm from Aberdeen, returning home with the 12.30am from Perth, the Aberdeen portion of the Corridor.

Up to the end of April 1896, time had been booked against a Dunalastair on only one occasion, two minutes having been lost on account of very severe weather with a train of 20 vehicles. Occasionally as much as 20 minutes of lost time were made up and loads of from 19 to 22 vehicles were frequently mastered without the aid of a pilot. From descriptions of footplate trips made by Rous-Marten, Lord Monkswell and others, it would seem that the engines did much of their work with the lever set in the second notch from mid-gear and with the regulator one-third to one-half open. With the down West Coast Postal, 135 tons full, No 726 was able to maintain around 68mph on level track with the regulator under half open and the lever in the first notch.

A detailed account of the perfomance of the Dunalastairs may be found in O. S. Nock's monograph *The Caledonian Dunalastairs* and, in the present study, it will be necessary only to summarise their achievements. Their finest work was done on the down Tourist which generally weighed 160-200 tons tare. Gains of six or seven minutes on schedule became the order of the day and when No 728, with inferior coal and 200 tons behind the tender, gained only two minutes, Rous-Marten classed her performance as a 'failure'. On another occasion, the same engine covered the first 100 miles in 97min but a 12-minute signal stop at Larbert spoiled what might have been a remarkable record. Rous-Marten estimated her net time to Stirling as 113min.

In considering such work as this, Lord Monkswell wrote in *The Railway Magazine* of April 1936:—

> It is, I believe, true to say that, when all the circumstances are allowed for, there is not, and never has been, any locomotive work regularly performed on a British railway to equal what the Caledonian engines did from July to November 1896. Even now there is not in this country (nor, so far as I know, in any other part of the world either) any train timed at 60mph start-to-stop over any lines as difficult as the section from Carlisle to Stirling.

CHAPTER 6

THE McINTOSH STANDARD CLASSES 1897-1900

THE NEW ERA

Although conforming to Drummond's basic precepts, McIntosh's standard tender engines represented a greater departure from existing practice than the quasi-Drummond Dunalastairs and introduced an elegance of line which set them apart from their forebears. The ultramarine of former years gave way to a lighter, brighter and more striking shade of blue. This was introduced in 1897 when Nos 723 and 724, named *Victoria* and *Jubilee*, were repainted to commemorate Queen Victoria's Diamond Jubilee. They were lined in gilt and royal purple. Later that year, No 721 was also repainted light blue, though with standard black and white lining. Several of the 'Coast Bogies' were painted dark blue with yellow lining but this style was not perpetuated.

The year 1897 brought changes in personnel. William Urie succeeded R. W. Urie as works manager in December, his position as chief draughtsman passing to Thomas Weir.

THE 766 CLASS

The Dunalastairs had proved exceptionally free-steaming engines and McIntosh must have felt that an increase in the diameter of the cylinders would not tax the ability of the boiler to produce the necessary steam. Since new and more powerful engines were needed to cope with the ever-increasing tonnage of the new corridor trains, 15 4-4-0s with 19in x 26in cylinders and a lengthened version of the Dunalastair boiler were constructed at St Rollox.

THE MCINTOSH STANDARD CLASSES 1897-1900

Order No	Engine Nos	Dates delivered	Cost per engine
Y51	766-80	12/97—4/98	£2434

The frames were 1ft longer than those of the earlier engines, the increase being in the section between the bogie-centre and the driving axle. The motion-plate again supported the rear of the slidebars but the longer frames allowed the eccentric-rods to be lengthened by 11in. Clearance for the bogie was provided by

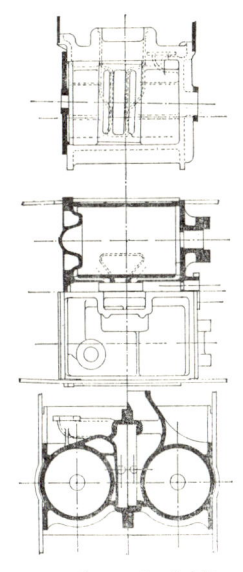

Fig 8　Arrangement of standard 19in × 26in cylinders

setting in the frames forward of the motion-plate and by inclining them towards each other forward of the cylinders. The Drummond front-end, with its steam-jacketed cylinders, divided ports and cone-shaped pistons, was abandoned: The whole of the exhaust steam went directly to the base of the blastpipe. Steam and exhaust port areas were increased to meet the greater cylinder volume and the shape of the cylinder covers changed to match the more conventionally shaped pistons.

Enlargement of the ports meant that larger slide-valves were required, tail-rods being fitted to provide the extra support and guidance required. Enlargement of the cylinders posed problems of steam-chest space and, to provide adequate room for the slide-valves, the cylinders were set with their centres $1\frac{1}{2}$in further apart than in the Dunalastairs, this requiring a slight bulging of the frames. The altered cylinder arrangement called for changes in the proportions of the crank-axle, the outer crank-webs being narrower than before and the journals projecting further through the frames, with consequent reduction in the depth of the wheel-seats. To meet the greater forces generated by the 19in cylinders, the diameter of the crank-pin and journals was increased by $\frac{1}{2}$in. The weighshaft and reach-rod were positioned above, instead of below, the eccentrics and elliptical crank-webs replaced the parallel-sided pattern formerly used. Concave journals providing a considerably greater bearing surface than the old straight journals were employed for the trailing coupled axle.

The firebox was identical to that of the Dunalastairs but the boiler-barrel was lengthened by $9\frac{1}{2}$in and the smokebox by 5in, while working pressure was increased from 160 to 175lb/sq in. The tubeplate arrangement was unchanged though the tubes were copper. The dome was enlarged but since its cover was only $1\frac{1}{4}$in taller and $1\frac{1}{2}$in wider than before, its increased size was not readily apparent. The chimney and liner were unchanged but the vortex blastpipe was replaced by a plain blastpipe, its $5\frac{1}{4}$in diameter orifice being situated $7\frac{1}{4}$in above the boiler centre-line.

Gresham and Craven steam sanding apparatus was fitted for forward running. Also new was the brass displacement lubricator, with T-shaped tap, attached to the front of the valve-chest, beneath the inspection cover at the base of the smokebox. This was supplemented by the sight-feed lubricator in the top left-hand corner of the cab. New additions within the cab were the steam sanding valve, situated on the faceplate to the left of the left-hand injector, and the carriage warming apparatus, the pressure-reducing valve being attached to the faceplate below the regulator handle. In other respects, the footplate resembled that of the Dunalastairs though the reversing quadrant had only six notches on either side of mid-gear. In both classes, the driver's brake valve was attached to the cab side, the Gresham and Craven No 9 combination injectors occupying all the available space on the faceplate. The brake system was refined by the incorporation of a governor for the Westinghouse pump.

The cab was widened by 8in, this necessitating an increase of 4in in the width of the splashers. The small splashers covering the

THE MCINTOSH STANDARD CLASSES 1897-1900

crank-pins were correspondingly narrowed and the long, low splasher connecting them omitted. The leading splashers were separate from the sandboxes, which were repositioned below the running-plate and partly hidden from view by the front footsteps. The appearance of the cab was improved by terminating the 'lean-out' 6½in short of the roof and the whole ensemble was perfected by the introduction of a handsome bogie tender which complemented the long, racy look of the engine. The tank was 2ft 6in longer, 5in wider but 3in shallower than the Lambie 3570-gallon tender and rested on double frames, the inside pair forming the outer walls of the well. Two strong castings (bolsters) connecting the outside frames provided the bearing surfaces for the bogies. The bogie frames were united at the ends and joined in the middle by the bogie-bolster, with its centre-pin and circular bearing surface 18in in diameter. Between the two axleboxes on each side was a single inverted laminated spring, fixed to the bogie-frame by a strong pin passing through the buckle. The ends of the spring were linked to a compensating beam which rested on the axleboxes. Since lateral movement of the bogies was severely limited, a single brake-cylinder, suspended below the footplate, sufficed for all eight wheels.

The first engines were no sooner out of St Rollox than Rous-Marten was in Glasgow where, under the expert eye of Driver Andrew Dunn, the first two engines were being tested on normal services. Rous-Marten's first run was made on the footplate of No 766 and involved a trip to Carlisle and back with the 'Corridor'. On the down journey, the load was 330 tons tare but despite the heavy load and a strong westerly wind, the engine mastered the 135min schedule. Minimum speeds up the 1 in 99 gradients to Craigenhill and Beattock summits were 30 and 31mph respectively but on favourable stretches the engine was run under very easy steam and speed was not allowed to exceed 75mph. On the return journey, with the load 276 tons and greasy rails an additional hazard, the train passed Rockcliffe (4.1 miles) in under 6min and Lockerbie (25¾ miles) in 30min 28sec. The favourable stretch from Beattock Summit to Strawfrank Jct was run at an average speed of 67mph with a maximum of 80mph. There was no doubt as to the engine's fleetness of foot. On the following day, No 767 took the 4.13pm Glasgow Central-Gourock boat train and, with a load of 200 tons tare, covered the 26¼ miles in 31min 44sec start to stop. This was another brisk performance for the route runs through congested built-up areas. The seven miles out to Paisley were taken very easily but on the slightly favourable stretch through Langbank 76mph was attained. In summarising his experiences, Rous-Marten commented:—

> Although these engines have hardly yet got into regular working, even their trial performances have proved them capable of realising the fullest intentions of their able designer..... Their appearance at the head of the train at Central Station, Glasgow, is always the signal for a large crowd to assemble; indeed on each occasion when I was present the engine was absolutely mobbed by admiring spectators. This is hardly surprising in consideration of the severe duty they are called on to perform, undertaking, unaided, over exceptionally severe grades, loads which on many lines, very much easier in character, would usually be taken by two engines or run in two portions. . . . Travelling on the footplate, I found the new engines ran with an ease and steadiness which left nothing to be desired. The boiler made steam very freely, keeping the needle constantly at blowing-off point.

Two days before the above appeared in print, a full-scale indicator test was carried out with No 772 on the 'Corridor', this time loaded to 305 tons tare. The average speed for the whole journey was 46mph, coal consumption being 49.6lb and water consumption 34.3 gallons per mile. Since the mean power developed in the cylinders throughout the run was 810ihp, coal and water rates per ihp-hour were 2.8lb and 19.5lb respectively. The excellence of the latter—for a saturated engine—spoke well of the design of the ports, valves and cylinders. The coal rate was slightly high but this was at least partly due to its indifferent quality, only 6.95lb of water being evaporated per lb of coal. (With coal-mines on its doorstep, the Caledonian used local Scottish coal, which was inferior to the best quality Welsh or Yorkshire coal used south of the border.) The efficiency of the boiler is shown by the satisfactory rate of evaporation—15,770lb per hour. Coal per gross ton mile was 0.122lb, a figure comparable to those established on tests with other saturated classes, such as Midland singles and LNWR Precursors. E. C. Poultney estimated that the maximum power output which could be expected from an engine of the class was about 1020ihp. An average of 940ihp was developed by a Dunalastair when hauling the 1896 down 'Tourist' between Carlisle and Beattock Summit.

THE MCINTOSH STANDARD CLASSES 1897-1900

The original allocation of the class was:—
  Polmadie   Nos 766-7   Edinburgh Nos 768-9
  Carlisle   Nos 770-4   Perth   Nos 775-80

Nos 766-71 and 773-4 were painted in the usual dark blue livery and the remainder in the light blue livery of the three named engines of the 721 class. It was reported in *The Locomotive* of July 1898 that 'one of these already famous engines, No 779, has the name *Breadalbane* painted on the driving splashers'. The August issue reported that No 766 had been repainted light blue and named *Dunalastair 2nd*. The tenders of all the engines carried a gilt scroll on either side of the coat of arms.

THE 812 CLASS

Following the successful debut of the 766 class, McIntosh produced a large-boilered 0-6-0. He was determined that it should retain the versatility of its predecessors for its official title was '18½in x 26in by 5ft passenger goods engine'.

Although derived from the later 'Jumbos', the new 0-6-0s owed much to the 721 and 766 classes. The cylinders and valve-chests were modelled on those of the 766 class, the steam and exhaust port areas being the same despite the ½in reduction in cylinder diameter. The wheelbase was 6in longer than that of the 'Jumbos', the 3in increase in the spacing of the leading and driving axles allowing the eccentrics to be lengthened from 4ft 7in to 4ft 11in, though the 6ft 6in connecting-rods were retained. Apart from a slight extension of the crank-webs, the crank-axle was identical to that of the 766 class. The boiler and firebox were of the type fitted to the 721 class but with repositioned dome (766 class pattern) and safety-valves. The draughting arrangements were similar to those of the 721 class but the smokebox was 2¼in longer and the chimney ½in shorter. This was the last occasion on which the vortex blast-pipe was used on a tender engine.

The cab was of the 766 class pattern but shortened by 6in. Boiler feed was by two Gresham and Craven No 9 combination injectors, and reverse by lever, the quadrant having six notches on either side of mid-gear. The brake-rigging was similar to that of the later 'Jumbos' and was operated by a 9in diameter steam or 15½in diameter compressed air cylinder slung vertically below the footplate. Sanding was by gravity and fore-end lubrication by a vacuum sight-feed lubricator in the left-hand corner of the cab,

THE MCINTOSH STANDARD CLASSES 1897-1900

two Furness lubricators on the smoke-box front and a single displacement lubricator on the front of the valve-chests. A new type of tender carrying 3000 gallons was derived from the Lambie 3570-gallon pattern, the sole difference being in the depth of the tank.

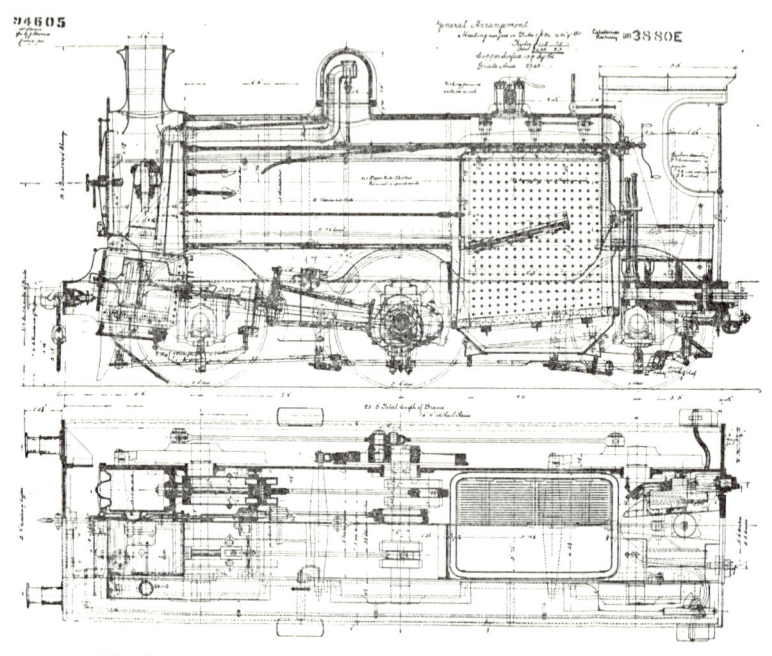

Fig 9 General arrangement of McIntosh standard 0-6-0

Two batches were built at St Rollox:—

Order No	Engine Nos	Dates delivered	Cost per engine
Y54	812-28	5/99—8/99	£2274
Y58	282-93	9/99—11/99	£2189

Batch Y54 were Westinghouse-fitted and painted blue. The remainder had steam brakes and were painted black though Nos 282-3 were fitted with vacuum ejectors at the end of 1904. Of the 17 passenger engines, Nos 812, 815, 817 and 826 are thought to have gone to Polmadie and Nos 813-4 and 818 to Greenock. Nos 821-2 appear to have been allocated to Edinburgh and Nos 823 and 825 to Dundee. Perth received No 824 and Aberdeen No 828. The Polmadie and Greenock engines worked from Glasgow to the

THE MCINTOSH STANDARD CLASSES 1897-1900

Clyde coast, their high adhesion and boiler power making them ideal for the steep banks of the Wemyss Bay branch. The Edinburgh engines ran the fast fish trains from Granton to Carlisle and worked excursions during holiday periods. No 828 was employed on the Aberdeen—Stonehaven—Laurencekirk passenger services but also ran excursions from Aberdeen to Edzell on summer Saturdays.

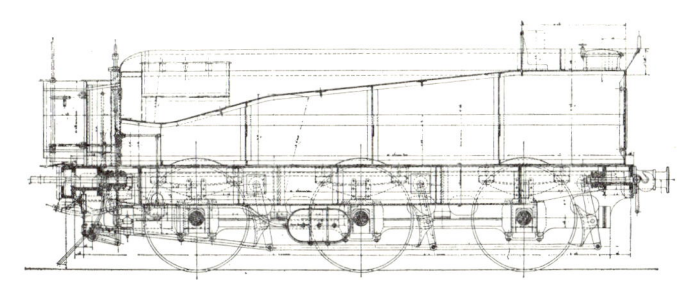

Fig 10 General arrangement of standard 3000 gallon tender

The turn of the century saw a boom in the British coal-mining industry and every railway serving the coalfields found itself short of suitable engine power. While several of the English companies filled the gap by importing American 2-6-0s, the Caledonian expanded its stud of 18½in 0-6-0s. St Rollox was then committed to a heavy programme of construction and, for the first time in 15 years, bulk orders for 50 engines were placed with private builders.

Builder	Neilson Reid & Co	Sharp Stewart & Co	Dübs & Co
Engine Nos	829-48	849-63	864-78
Works Nos	5613-32	4634-48	3880-94
Dates delivered	12/99—7/00	8/00—9/00	4/00—5/00
Cost per engine	£3100	£2985	£2985

The 50 engines were fitted with the steam brake and painted black. They were unusual in having the Drummond numberplate and three-link couplings; the St Rollox-built engines had standard McIntosh numberplates and single-link couplings. Although intended for goods work, these engines were given a superb finish. First, the whole of the engine was cleaned with turpentine and German rubbing-down stone. One coat each of best red lead and lead colour oil paint were then applied to the smokebox, boiler clothing, cab, splashers, buffer beams, running-plate, frames (outer surface) and tender sides and frames. Any irregularities in these surfaces were then filled with white lead putty and the whole given a coat of patent 'filling-up'. After this had dried and been rubbed down, another coat of lead colour was applied. The frames were then ready for painting and varnishing but the other surfaces mentioned above were worked over again with rubbing-stone and, except for the smokebox, given a final coat of lead colour. Only then was the black livery applied in two coats, the boiler, cab, splashers and tender sides being rubbed with glass-paper between each. Following a coat of varnish, the engine was lined out and the whole process completed by the application of two further coats of the best varnish. The buffer beams received two coats of vermilion and two of varnish. The inner surfaces of the frames were filled up and rubbed down, given one coat of primer, rubbed

TABLE 4
812 CLASS WORKINGS

Train	7.10pm Carlisle-Polmadie			1.45am Carlisle-Greenock		
Date	11/3/19			4/1/17		
Engine No	832			834		
	arrive	leave	load*	arrive	leave	load*
Carlisle No 3 Box		7.20pm	32		2.8am	36
Kirkpatrick	7.50	8.05	32			
Kirtlebridge	8.15	8.28	32			
Ecclefechan	8.35	8.45	32			
Murthat	9.30	9.48	32	Signal stop. 10 minutes		
Beattock	9.58	10.42	32	3.44	3.55	36
Auchencastle	10.47	10.57	36			
Symington				?	5.30	
Carstairs	12.35am	12.50am	36	5.50	6.00	32
Wishaw	1.33	1.40	36			
Bellshill	2.00	2.05	36	7.10	7.15	32
Viewpark Box	2.10	2.15	36	7.25	8.20	32
Carmyle	2.25	3.05	6			
Rutherglen				8.33	8.55	36
Polmadie	3.35	—				
West Street				9.15	?	36
Paisley Goods				9.30	10.00	36
Port Glasgow				10.30	10.40	10
Greenock				10.55	—	

Banked by No 952   Banked by No 950
Beattock-Summit   Beattock-Summit

* Including brakevan

Page 87 (upper) 600 class 0-8-0 No 604; *(lower)* 652 class 0-6-0 No 423

Page 88 (*upper*) 55 class 4–6–0 No 59 at Stirling; (*lower*) 918 class 4–6–0 No 919 at Carlisle

THE MCINTOSH STANDARD CLASSES 1897-1900

down again with glass-paper and finished with two coats of vermilion and two of varnish. The inside of the cab was similarly prepared and, following two coats of oak colour, was grained and finished with two coats of varnish. The wheels got two coats of lead colour, one of black and three of varnish.

The steam-braked 18½in 0-6-0s were mainly employed on longer-distance work, the *approximate* numbers allocated to each shed at any one time being as follows:—

Carlisle	13-16	Carstairs	4-5	Edinburgh	9
Greenock	3	St Rollox	8-10		
Perth	7-9	Dundee	4-5	Aberdeen	3-5

An idea of their work is given in Table 4. This has been prepared from guards' journals and, while not absolutely precise as to times, provides reasonably accurate logs of the journeys. A considerable amount of time was spent either waiting in loops or taking on water. For such work, the 18½in 0-6-0s were well proportioned, combining pulling power with economy.

THE 900 CLASS

An improved version of the 766 class was built at St Rollox between December 1899 and July 1900:—

Order No	Engine Nos	Date delivered	Cost per engine
Y57	900-02	12/99	£3049
Y62	887-99*	4/00—7/00	£3049

* Not finished in numerical order

These engines differed from their predecessors in several respects. The firebox was lengthened by 6in and deepened by 3in, by increasing the coupled wheelbase by 6in and raising the boiler centre-line by 3in. The shortening of the chimney which resulted required modification of the draughting, the blastpipe orifice being 7¼in lower, relative to the boiler centre-line, and ¼in greater in diameter than before, while, projecting downwards from a point 4in within the hood, was a conical petticoat pipe. The number of tubes was increased by four, while working pressure was raised to 180lb/sq in and the firebox crown strengthened; in the 721 and 766 classes, the longitudinal girder stays had been suspended from two angle-irons riveted to the roof of the outer firebox but, in the new engines, two pairs of angle-irons were employed, the two members of each pair being linked by a stout bar.

THE McINTOSH STANDARD CLASSES 1897-1900

The lower arm of the regulator handle was lengthened to compensate for the increased height of the faceplate. Steam reversing gear was provided. The manually-operated lever was retained but was coupled to the piston of a steam cylinder just above the floor of the footplate. To alter the cut-off, the catch of the lever was released from the notch in the reversing quadrant and the steam turned on, the lever being moved backwards or forwards as required. Speed of movement was controlled by an oil cylinder, this acting solely as a brake; movement was stopped by re-engaging the catch in the appropriate notch of the quadrant. If desired, the lever could be operated manually, this being an advantage when the engine was not in steam. In previous classes it had been virtually impossible to draw the lever back while the regulator was open but the very opposite problem now applied, for, in the absence of a proper hydraulic brake, the heavy lever could move with lightning speed and inflict serious injuries on the driver. Thus, despite an increase to nine notches on either side of mid-gear, expansive working was positively discouraged.

The cab lean-out was altered to a curved form which suited the lines of the engine much more than the rather square-cut form used on the 766 class. A less noticeable change lay in the shape of the rear angle-iron supporting the cab roof; this was most obvious when the engine was viewed from above, the rear of the cab appearing straight instead of curved.

The tender coping was raised to match the higher pitched boiler. The tender bogies were re-positioned, the leading overhang was lengthened, the trailing overhang shortened and the bogie-centres brought 3in closer together. In April 1900 the fitting of vacuum ejectors and pipes to the 13 engines of batch Y62 was authorised at a cost of £50 per engine.

The 900 class soon acquired the unofficial title 'Dunalastair III' class; no engine actually bore this name. The probable allocations were:—

Polmadie	Nos 900-2	Edinburgh	Nos 896, 899
Carlisle	Nos 893-5, 897-8	Perth	Nos 887-92

The three Polmadie engines went to drivers Ranochan, Currie and Dunn and were employed on the 'Corridor' and on certain of the Glasgow-London night services. The up 'Corridor' was generally composed of 'equal to 21½' vehicles, a tare weight of over 360 tons, for which a pilot was often provided to Carstairs. The

THE MCINTOSH STANDARD CLASSES 1897-1900

load of the down train was 'equal to 16½' vehicles, a tare weight of just over 300 tons. In describing this task, *Engineering* commented:—

> ... and this train has been worked from Carlisle to Beattock without assistance, a distance of 39¾ miles, in 42½ minutes, which, in view of the gradients, and of the fact that from Carlisle to Beattock the line rises to a height of 350ft, must be pronounced exceptionally good work.

The 900 class went into revenue-earning service without the publicity which had ushered in their predecessors. For five years they shouldered the hardest tasks which the traffic department could give them. When enumerating the best British runs of 1899, Rous-Marten wrote:—

> One run on the Caledonian at 59 miles an hour; two on the same line at over 56; three on the same line over 55. Those are in Scotland. In England proper, we have not a single run at 60 or 59, or 58 or 57 or 56 miles an hour. No, not one.

Rous-Marten and his friends paid less attention to the work of the 900 class than they had done to that of the 721 and 766 classes and the maximum power output was never established. Of the few detailed recordings which exist, one made by Lord Monkswell in May 1902 illustrates the capabilities of the class. The engine was No 902, with 320 tons gross on the down 'Corridor'. Once the train was on the move, the reversing lever was put into the third notch (about 32 per cent cut-off) and with full regulator a rapid start was made, the 4.1 miles to Rockcliffe being run in 6min 36sec. Thereafter, the engine was driven entirely on the regulator, the opening varying from one-half to two-thirds full. This provided a maximum speed of 62mph at the Solway Firth and 72 after Lockerbie, the intervening uphill section being covered at about 50mph. Beattock was reached in 44min 43sec, 2¼min slower than in the run referred to by *Engineering*. There the mandatory pilot was coupled to the rear and a vigorous asault was made on the bank, the summit being passed in 18½min. In describing the hard collar work, Lord Monkswell wrote:—

> The engine had run remarkably smoothly and was nowhere in the slightest difficulty. The fire appeared to be rather thin, well below the firehole door behind where all the fresh fuel was put on. The firing was more or less continuous but all done in the most leisurely way,

THE MCINTOSH STANDARD CLASSES 1897-1900

the fireman raising the deflector plate with his right hand just enough to get in the shovel with his left, and dropping the coal under the firehole.

STANDARD TANK ENGINES

To provide cheap but adequate power for the company's short-distance mineral and passenger trains, McIntosh introduced three classes of non-condensing tank, each fitted with the latest version of the 'rebuild' boiler. The first of these, the 782 class, was a 0-6-0T with 4ft 6in wheels and standard 18 by 26in cylinders. Basically a non-condensing version of the 29 class with Gresham and Craven combination injectors and, in most cases, a steam brake only, it was ideally proportioned for short-distance mineral and yard duties and, numerically, became the second largest class on the system. Details of construction were as follows:—

Order No	Engine Nos	Dates delivered	Cost per engine
Y52	782-8	6/98—7/98	£1542
Y53	789-811	8/98—12/98	£1569
Y55	236-47, 485, 516	12/98—1/99	£1569

Most of these engines were sent to Motherwell and Hamilton sheds but Polmadie got five and Edinburgh, Carstairs and St Rollox two or three each.

McIntosh next turned his attention to the provision of suitable locomotives for working passenger trains on the Cathcart Circle, opened in April 1894, and on the Balerno branch. Since both lines were sharply curved, with steep gradients and stations spaced at frequent intervals, a small-wheeled 0-4-4T was considered the most suitable type:—

Order No	Engine Nos	Dates delivered	Cost per engine
Y56	104-11, 167-70	3/99—6/99	£1719

The valve-chests and 17in by 24in cylinders were closely based on those of the 1 class, the ports being undivided. The slidebars and motion-plate were arranged on the McIntosh plan and the crank-webs were of the new elliptical pattern. The coupled wheels were 4ft 6in in diameter. The boiler was standard with that of the 782 class and was fed by two combination injectors but a relic from the past was the vortex blastpipe exhausting into a liner-less chimney.

THE MCINTOSH STANDARD CLASSES 1897-1900

A great improvement on that of all previous side-tank engines was the enlarged cab, the side-sheets being lengthened by 8in and made flush with the tanks, the overall width of the cab thus being increased from 6ft 8in to 7ft 9in. The sight-feed lubricator was in the front left-hand corner of the cab and the quadrant carrying the levers operating the cylinder cocks and the forward and backward sanding lay on top of a toolbox to the left of the reversing lever. The Westinghouse brake valve was attached to the side-sheet above the toolbox and the Westinghouse steam cock to the extreme right of the faceplate.

The suspension and brake rigging arrangements were similar to those of the 19 and 92 classes but, because of the wide cab, the Westinghouse pump was to the front of the right-hand side-tank, while the reservoir was slung transversely below the rear of the bunker. The sandboxes for backward running were attached to the main frames to the rear of the driving wheels. Another innovation was the fitting of a governor to the Westinghouse pump, though this had already appeared on tender engines commencing with the 766 class. The bogie was similar to that of the Drummond 0-4-4Ts, with the same 5ft wheelbase and 2ft 6in wheels, the latter, however, having eight spokes. Water capacity was 1000 gallons and coal 2¼ tons.

The engines were painted blue. Nos 104-9 and 167-70 went to Edinburgh for the Balerno branch while the remainder were based on Polmadie for the Cathcart circle, though changes were made from time to time. After World War I, No 104 was carriage-pilot at Buchanan Street while about 1921 No 110 worked the Beith branch.

For suburban and branch line passenger work, McIntosh produced a non-condensing version of the 879 class 0-4-4T:—

Order No	Engine Nos	Dates delivered	Cost per engine
Y61	439-43	3/00—4/00	£2031
Y64	444-55	9/00—12/00	£2151

Combination injectors were fitted and working pressure was increased to 160lb/sq in. Some minor experimentation in draughting seems to have been made on the first batch, for the blastpipe and chimney drawings, though stamped Y61, carried the instruction 'For 3 engines only'. Water capacity was 1270 gallons and coal 2½ tons.

THE MCINTOSH STANDARD CLASSES 1897-1900

The engines were painted blue. Three or four were sent to Beattock and a similar number to Perth and its sub-sheds. The Forfar area got one or two, the remainder going to Polmadie and the main Lanarkshire sheds.

Another eight 0-4-0STs were constructed:—

Order No	Engine Nos	Dates delivered	Cost per engine
Y63	621-6	7/00—8/00	£1079
Y68	627-8		

These engines had springs in a prominent position above the running-plate and standard McIntosh numberplates but were otherwise similar to those of batch Y43.

CHAPTER 7

McINTOSH DEVELOPMENTS 1901-4

THE 600 CLASS 0-8-0s

The turn of the century saw the introduction of several types of high-capacity bogie wagon for mineral traffic. In October 1900, approval was given for the purchase of 50 30-tonners, and in December 1901 the board authorised the construction of 500 for the conveyance of locomotive coal. With so many high-capacity Westinghouse-braked wagons on order, McIntosh designed an eight-coupled mineral engine, with the Westinghouse brake. His decision was made with some reservation for the scarcity of long sidings and loops limited the length of the trains to 15 of the new wagons, a tonnage which could be handled satisfactorily by the much cheaper 812 class engines. Only eight 0-8-0s were constructed:—

Order No	Engine Nos	Dates delivered	Cost per engine
Y65	600-1	7/01	£3000
Y67	602-3	1/03-2/03	£3000
Y70	604-7	5/03-6/03	£3200

The 21in by 26in cylinders were inclined at 1 in 10 with the steam chests on top. As Stephenson link motion was retained, the slide-valves had to be driven indirectly via rocking levers and, to accommodate the standard 4ft 7in eccentrics, the driving axle was placed 8ft 6in behind the leading axle. Furthermore, to provide room for a well-proportioned firebox and unrestricted ashpan, the distance between the intermediate and trailing axles was also fixed at 8ft 6in. The driving and intermediate axles were only

MCINTOSH DEVELOPMENTS 1901-4

TABLE 5
600 CLASS VALVE SETTING: FORWARD GEAR. 1910 SETTING

Notch	Travel, in	Lead, in FP	Lead, in BP	Port opening, in FP	Port opening, in BP	Cut off, % FP	Cut off, % BP	Release, % FP	Release, % BP
8	$5\frac{3}{16}$	0	$\frac{3}{16}$	$1\frac{11}{16}$	$1\frac{1}{2}$	88	83	96	95
7	$4\frac{11}{16}$	$\frac{1}{16}$	$\frac{1}{4}$	$1\frac{7}{16}$	$1\frac{1}{4}$	84	$77\frac{1}{2}$	$94\frac{1}{4}$	93
6	$4\frac{3}{16}$	$\frac{3}{32}$	$\frac{11}{32}$	$1\frac{1}{8}$	$1\frac{1}{16}$	80	71	93	91
5	$3\frac{3}{4}$	$\frac{1}{8}$F	$\frac{3}{8}$	$\frac{7}{8}$	$\frac{7}{8}$	75	63	90	87
4	$3\frac{3}{8}$	$\frac{3}{16}$	$\frac{13}{32}$	$\frac{11}{16}$	$\frac{11}{16}$	66	51	87	$81\frac{1}{2}$
3	$3\frac{1}{16}$	$\frac{3}{16}$F	$\frac{13}{32}$	$\frac{15}{32}$	$\frac{19}{32}$	55	39	82	74
2	$2\frac{13}{16}$	$\frac{7}{32}$	$\frac{1}{2}$B	$\frac{5}{16}$	$\frac{1}{2}$	41	27	76	65
1	$2\frac{23}{32}$	$\frac{7}{32}$	$\frac{1}{2}$	$\frac{1}{4}$	$\frac{15}{32}$	24	$17\frac{1}{2}$	65	52

FP=front port FB=back port
F=full B=bare
Lap of valve 1in

5ft 4in apart, giving the wheelbase an asymmetrical appearance. The $1\frac{1}{8}$in thick frames were strongly cross-stayed between the driving and the intermediate axles. Behind the motion-plate was an additional cross-stay from which the intermediate valve-spindles were suspended.

The journals of the driving axle were of the same diameter as those of the 900 class but $\frac{1}{2}$in greater in length. The driving wheels were flangeless and the coupling rods jointed at the driving and intermediate wheels, the intermediate axle having a side-play of $\frac{1}{2}$in and the trailing axle one of $1\frac{1}{4}$in. Underhung laminated springs were used for the six leading wheels, the leading and driving on each side being equalised by a large inverted laminated spring. The trailing wheels were sprung by two spiral springs above the axleboxes, these acting against a plate running across the floor of the cab, the height of the plate and hence the tension on the springs being governed by a large nut above either end.

The boiler was a lengthened version of the 900 class boiler, the barrel consisting of three telescopic rings containing 275 $1\frac{3}{4}$in tubes and pitched at the same height as in that class. The firebox and dome were unchanged but the Ramsbottom safety-valves were replaced by four 4in lock-up valves in a cylindrical casing. Working pressure was 175lb/sq in. The chimney liner, hood and petticoat were of the type used on the 900 class but the smokebox was lengthened by 7in—this contributing to the 'front heavy' appearance of the engines—and the centre-line of the blastpipe orifice was placed $\frac{1}{16}$in in front of the centre-line of the chimney. Sanding was by gravity and reverse by unassisted lever, the quad-

MCINTOSH DEVELOPMENTS 1901-4

rant having eight notches on either side of mid-gear. The cab was similar to that of the 900 class, though 3in longer, while the tender was of the Lambie 3570-gallon pattern, but with a wider platform and with the addition of a water-level gauge. Although Westinghouse-fitted, all eight engines were painted black but had the beautiful finish of the express engines.

Most of the class were re-allocated several times. In September 1901 *The Locomotive* reported that Nos 600-1 were running on the 'southern and Ardrossan sections'. They were apparently being used on export coal traffic from the Lanarkshire coalfield to Ardrossan harbour. In July 1903, *The Locomotive* reported that No 602 had run trials between Motherwell and Perth with 30 of the 30-ton bogie wagons. Later that year, during the Glasgow Fair holidays, several of the engines worked special boat trains from Glasgow to Ardrossan. On 24 July, No 604 was derailed at Arkleston Jct, near Paisley, two steam cranes and a large squad of men working well into the following morning before the engine was re-railed. In 1905, Greenock, Stirling, Perth and Dundee each had one engine, the remainder being at Polmadie, Motherwell and Hamilton. The Greenock engine worked to Dundee with the 8pm on Sundays and the 8.25pm on Tuesdays and Thursdays and returned with the 9.20pm on Mondays, Wednesdays and Fridays. The Dundee engine took the 11pm to Greenock on Sundays and alternated day about with the Greenock engine on the 8.25pm and 9.20pm services. The Stirling engine ran the 8pm to Dundee and the 1.45am Dundee to Plean, while the Perth engine took the 10.45am to Ross Jct and returned with the 3.10pm. The Motherwell engines ran over the Lanarkshire and Ayrshire line with the 2.25pm from Motherwell Colliery to Glengarnock. On the Lesmahagow New Line, the following limits, in numbers of four-wheeled mineral wagons, were laid down:—

		Class		
		600	812	Jumbo
Merryton Jct—Stonehouse Jct	Full	35	22	20
	Empty	60	45	40
Stonehouse Jct—Blackwood	Full	40	27	25
	Empty	60	50	45
Blackwood—Alton Heights	Full	35	22	20
	Empty	60	45	40

MCINTOSH DEVELOPMENTS 1901-4

During their earlier years, the Motherwell and Hamilton engines spent much of their time on the main line, a Hamilton 0-8-0 regularly crossing Beattock Summit with the 4.15am from Strathaven Jct to Lockerbie and the 10.25am return working. In later years, however, the class seldom strayed far from the mineral belt.

The 0-8-0s were disliked by the shed staff. Although a large front-end cover was provided, the slide-valves could be removed only by taking out the blastpipe, rigging up a block and tackle in the chimney and winching them up into the smokebox. The adjusting nuts for the trailing springs corroded, and renewal of the springs involved hours of work in getting the nuts broken off.

NEW BOILERS

After 1900, 10 of the older Drummond 4-4-0s were rebuilt with larger boilers. The first engine dealt with, No 71, left the works in July 1898 with a 721 class boiler and cab and McIntosh brake arrangement. With its double front footsteps opposite the rear bogie wheels, combined sandbox-splashers and Drummond numberplates, this engine looked every inch a Dunalastair I, though it was coupled to a Drummond 3550-gallon tender. An undated entry in the locomotive register gave the cylinder diameter as 18¾in. Between March 1901 and April 1902 nine engines were rebuilt, Nos 60, 62-6, 70, 73 and 75. All got cabs of 900 class pattern, curved splashers, sandboxes beneath the running-plate, single front footsteps opposite the sandboxes, McIntosh numberplates and Drummond 3550-gallon tenders. Dunalastair type boilers were fitted but most, if not all, of these were of the 812 class variant. Vacuum ejectors and through pipes were fitted to all nine engines, the steam being taken from a fountain beside the whistle and passed through a brass pipe into the cab.

Around 1906, Nos 70 and 75 were coupled to bogie tenders of 900 or 140 class pattern.

No 70 was shedded at Perth but No 75, which had been a Perth engine, was transferred to St Rollox and employed on the 7.5pm from Buchanan Street (Goods) to Carlisle and the 11pm return working. Apart from No 61, which had already been rebuilt with a Lambie boiler (see Chapter 5), the remainder of the Drummond 4-4-0s were rebuilt with McIntosh boilers designed in March, 1898, and similar to those of the McIntosh 'Jumbos'. The same

MCINTOSH DEVELOPMENTS 1901-4

boiler was also used in the rebuilding of some of the earlier 'Jumbos'.

In 1905, 'Coast Bogie' No 85 was rebuilt with a larger and higher-pitched boiler, the details being as follows:—

Boiler: maximum external diameter	4ft 6¼in
Boiler: pitch	7ft 6in
Length between tubeplates	10ft 4in
Length of outer firebox	5ft 5in
Heating surface: tubes	1047sq ft
Heating surface: firebox	111sq ft
Grate area	17sq ft
Working pressure	170lb/sq in
Adhesion weight	29 tons 2¾cwt
Weight in working order	43 tons 9¾cwt

The remainder of the class were rebuilt with the 1898 or 1904 version of the 'rebuild' boiler. The former version was also used in rebuilding Brittain's 'Oban bogies', Nos 179-88, between 1898 and 1901. Other Brittain engines which got the 'rebuild' boiler were the 30 670 class 0-4-2s of 1878/81, both varieties of boiler being used. These engines, by then numbered 670-9, 248-55, 275-6 and 278-87, were also fitted with the Westinghouse brake and transferred to local and branch-line passenger traffic.

THE 55 CLASS

The rebuilt Brittain 4-4-0s were small engines and it became obvious that they could not cope single-handed with the heaviest trains on the Callander & Oban. In 1901 thought was given to the production of an engine of increased adhesion and higher tractive effort but with a wheelbase suited to the sinuous nature of the track. In 1902 five small 4-6-0s were constructed at St Rollox, an additional four of slightly modified design following in 1905:—

Order No	Engine Nos	Dates delivered	Cost per engine
Y66	55-9	5/02—6/02	£2898
Y75	51-4	8/05—9/05	£2493

The coupled wheels were only 5ft in diameter and the distance between the driving and intermediate axles 5ft 3in, this being made possible by placing the brake-hangers in front of the driving wheels but behind the intermediate wheels. Similarly, by applying the

MCINTOSH DEVELOPMENTS 1901-4

brakes to the rear of the trailing wheels, which were positioned as close to the intermediate brake-hangers as possible, the coupled wheelbase was kept to 11ft 3in. The bogie was 9in shorter than that of the 4-4-0s, clearance being provided by cranking in the frames just behind the rear bogie wheels. The 19in by 26in cylinders, with slide-valves between them, were similar to those of the 766 and 900 classes but were inclined at 1 in 9, the pistons driving the leading coupled wheels. The crank-axle was standard with these classes and there was just enough room between it and the motion-plate for the accommodation of 6ft 6in connecting rods and 4ft 7in eccentric rods as used on the Dunalastairs and McIntosh Jumbos. Suspension was by laminated springs under each of the six coupled axleboxes.

The boiler-barrel was a lengthened version of the 900 class boiler and was pitched at the same height. It was composed of three rings arranged telescopically, with the smallest at the front and contained 275 1¾in steel tubes. The firebox was only 6ft 5in long and, due to the presence of the trailing axle, was severely limited in depth, its heating surface being less than that of any other tender engine since the 'Coast Bogies' of 1888. The trailing axle also restricted the size of the ashpan, the rear portion of which was very shallow; a damper was fitted at the front only. The safety-valves were of the type and size used on the 600 class and were housed in the same cylindrical casing. The hood and petticoat were arranged as in the 900 class but the centre-line of the blastpipe orifice was placed ½in ahead of the centre-line of the chimney. In batch Y75, the smokebox tubeplate was recessed 9in into the boiler-barrel.

Sanding was by gravity, this being the first instance in which sand-pipes were fitted to a tender engine for backward as well as for forward running. (These had previously been fitted to the front of the tender). The leading sand-boxes projected through the running-plate. Reverse was by steam-assisted lever, the reversing cylinders being bolted to the frames above the weighshaft. Front-end lubrication was similar to that of the 766, 812 and 900 classes but in batch Y75 the Furness lubricators were replaced by pear-shaped displacement lubricators. Tablet-exchanging apparatus was fixed to the side of the cab. The 3000-gallon tender was a truncated version of the Lambie 3570-gallon pattern, the wheelbase being reduced by 2ft and the length of the tank and frames by 2ft 11in.

MCINTOSH DEVELOPMENTS 1901-4

The slots in the frames were omitted. Coal-rails were fitted to the tenders of batch Y75.

The nine engines were painted blue, five going to Stirling and four to Oban shed. They were mostly confined to the Stirling-Oban section where they successfully worked the seven passenger trains (four in winter) and two goods trains daily in each direction.

THE 49 CLASS

By the summer of 1902 certain of the principal Anglo-Scottish expresses had increased to 380 tons and, between Glasgow Central and Carstairs and Edinburgh and Cobbinshaw, a pilot engine, either a rebuilt Conner 7ft 2in 2-4-0 or an unrebuilt Drummond 4-4-0, was frequently required. To eliminate double-heading and provide data on which future policy could be based, two large 4-6-0s were built at St Rollox:—

Order No	Engine Nos	Dates delivered	Cost per engine
Y69	49-50	3/03—4/03	£3000

On the basis of tractive effort, they were the most powerful express engines in the United Kingdom and, at 73 tons in working order, were also the heaviest.

The cylinders were 21 by 26in, spaced at 2ft 1½in centres with steamchests on top as in the 600 class but with redesigned steam and exhaust ports. Steam distribution was by plain unbalanced slide-valves driven indirectly via rocking levers pivoted from a forward extension of the motion-plate, the intermediate valve-spindles being suspended from pendulum links attached to an auxiliary motion-plate. As the rocking levers reversed the direction of the valve-spindles, each eccentric-sheave had to be moved 180° round the crank-axle in order to maintain outside admission; when viewed with the crank at its rear dead centre, the centre of the forward gear sheave lay below the centre of the axle while that of the backward gear sheave lay above it. But since the forward gear eccentric rod was still coupled to the top of the expansion link and the backward gear rod to the bottom, the eccentric rods appeared to cross each other. (The significance of this will be discussed in Chapter Ten.)

The pistons drove the leading coupled wheels and since these were closer to the bogie wheels than in the 900 class, the presence of rockers, etc., made for a rather cramped layout, eccentric rods

101

MCINTOSH DEVELOPMENTS 1901-4

and connecting rods being restricted to the non-standard lengths of 4ft 6in and 6ft 8in respectively. With the short connecting rods, there was considerable inequality in the cut-off during forward and backward strokes:—

Notch	Lead, in FP	BP	Port opening, in FP	BP	Cut-off, % FP	BP	Release, % FP	BP	Valve-travel, in
10	0	$\tfrac{3}{16}$	$1\tfrac{7}{16}$	$1\tfrac{1}{4}$	85	79	95	95	$4\tfrac{23}{32}$
5	$\tfrac{1}{8}$	$\tfrac{3}{8}$	$\tfrac{9}{16}$	$\tfrac{5}{8}$	61	46	85	80	$3\tfrac{3}{16}$

FP = front port   BP = back port

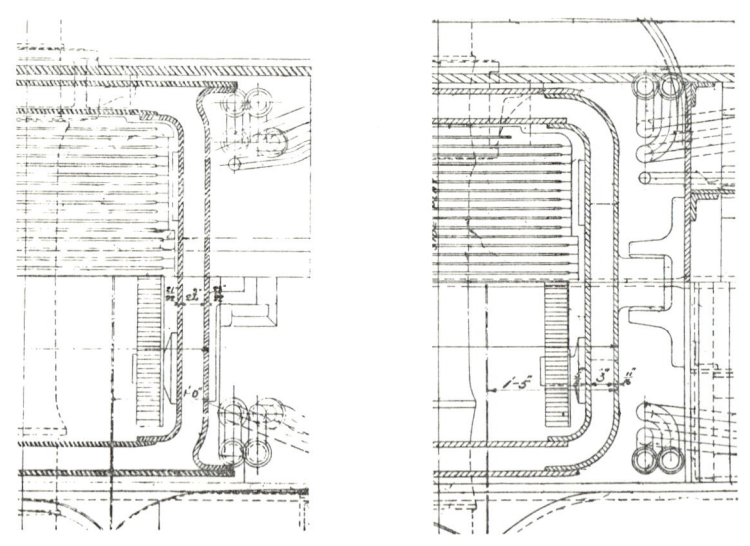

Fig 11   Arrangement of firebox faceplate: *Left* 49 class *Right* 903 class

The solid-forged crank-axle was a stouter version of the standard type, the journals being 1in greater in diameter and 2in longer than those of the 900 class. The frames were on a massive scale, 37ft long and 1⅛in thick and strongly cross-stayed between the driving and the intermediate coupled axles. Suspension was by laminated springs below all six coupled axleboxes. The bogie was pivoted 1in ahead of its centre-line, clearance being provided by cranking in the mainframes just behind the rear bogie wheels.

The three-ringed boiler-barrel was pitched 6in higher than that of the 900 class, the front and rear rings (both of 5ft external diameter) overlapping the ends of the intermediate ring. It con-

MCINTOSH DEVELOPMENTS 1901-4

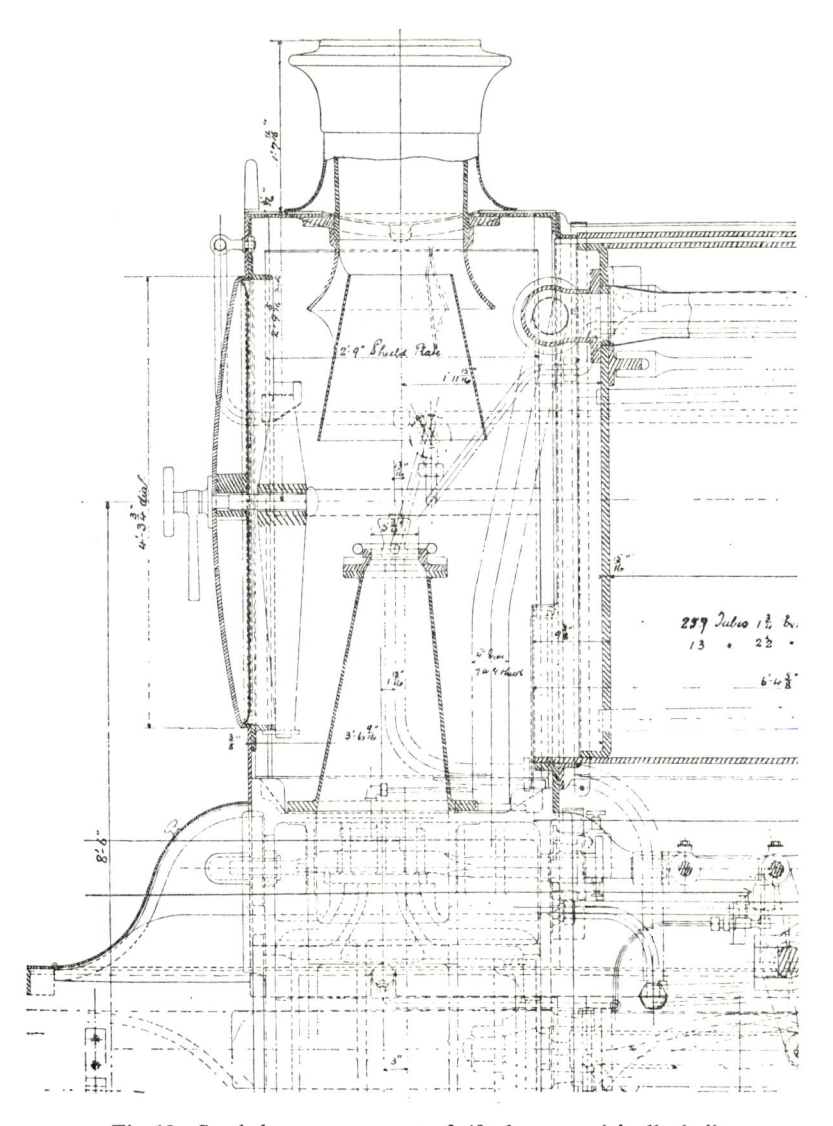

Fig 12   Smokebox arrangement of 49 class as originally built

MCINTOSH DEVELOPMENTS 1901-4

tained 257 1¾in copper and 13 2½in steel tubes, the latter being located at the bottom to provide additional exhaust space where char was most likely to accumulate. The smokebox tubeplate was recessed 9⅛in inside the barrel, length between tubeplates being 17ft 3in. The outer firebox was 8ft 6in long and provided room for a uniformly sloping grate of 26sq ft. The trailing axle restricted the depth of the firebox and ashpan, the latter being similar to that of the 55 class, i.e., with front damper only. The faceplate was riveted to the outer firebox casing in Crewe fashion, i.e., with its flanges pointing backward instead of forward (see diagram). Boiler feed was by two No 10 combination injectors. The four 4in 'lock-up' safety-valves were set to blow off at 200lb/sq in and were enclosed in a cylindrical brass casing. The smokebox arrangement was like that of the 900 class, but the centre-line of the blastpipe was placed 1⅜in ahead of the centre-line of the chimney.

Reverse was by steam-assisted lever, the reversing cylinders being located immediately above the footplate floor. The quadrant had 10 notches on either side of mid-gear and, for decorative effect, was covered by a notched brass plate. Sanding was by steam, but delivery was made to the intermediate as well as to the driving wheels. The sandboxes were below the running-plate. The splashers were of the narrow type, with the separate coupling-rod splashers favoured by Drummond. Since the cab was 4in wider than that of the 900 class, the continuity which had previously existed between the cab side-sheets and the rear splashers was lost. The brass spectacle frames were of a new form. Footplate fittings included a vacuum ejector and dual brake handle.

A wooden seat was provided for the driver and a larger folding one, presumably for an observer, against the right-hand side-sheet. Front end lubrication was on the standard McIntosh system but the displacement lubricator was attached to the rear of the valve-chests and the Furness lubricators to the inside of the frames to the rear of the cylinders. The tender was of the 900 class type but the tank was 4in wider and 10¼in deeper. The well was shortened and made up of three compartments, the middle one being much deeper than the others.

The emergence of such impressive locomotives aroused considerable interest among engineers and enthusiasts. Both engines ran trials while painted in works grey, No 49 being spotted at Edinburgh and No 50 at Perth. While No 50 was still under test,

Page 105 (upper) 49 class 4-6-0 No 50, *Sir James Thompson;* (*lower*) *Cardean* outside Crewe North shed, June 1909. Note lampirons above buffers for LNWR working and large diameter whistle

Page 106 (upper) 908 class 4–6–0 No 911, *Barochan*, on Gourock train at Glasgow Central; *(lower) Cardean* outside Polmadie shed. Note original whistle

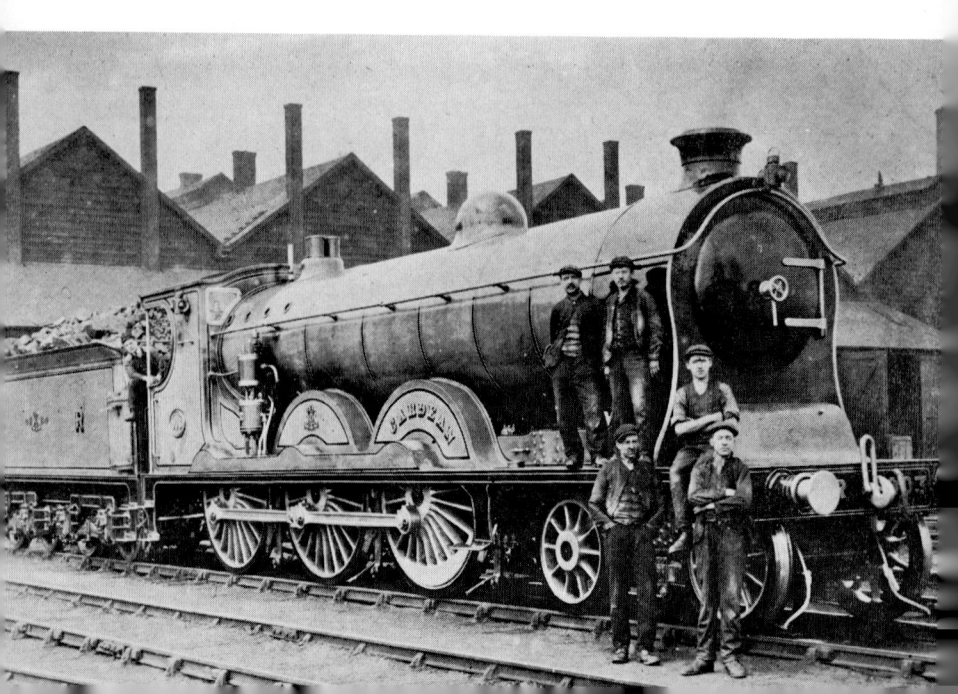

MCINTOSH DEVELOPMENTS 1901-4

No 49 was back in St Rollox being painted blue preparatory to taking over the 2pm up Corridor and the corresponding down express, the 8.21pm from Carlisle. When, in turn, No 50 entered the paint shop to receive her blue finery, the name *Sir James Thompson* was applied in gilt letters to her middle splashers. The finish was the best that St Rollox could provide, the finest camel-hair brushes being used to apply four coats of lead colour, two of blue, black or crimson and three of varnish. The tenders carried the company's initials in a much enlarged form and the coat of arms was surrounded by a wreath of gilt thistles and leaves, hand-painted by James Jeffrey, a St Rollox master craftsman. By the end of June the two engines were working the 2pm up Corridor and the 10.45pm up sleeper week about.

In an age when elegance counted almost as much as performance, the new engines provided eloquent testimony to the Caledonian's pre-eminence among the Scottish railways. A large oil-painting (now in the Glasgow Museum of Transport) of No 50 and train was commissioned and reproduced in postcards and other advertising material. The engines were also used on special excursion trains. When the employees of Dübs & Co held their summer outing in 1903, Nos 49 and 50 hauled them to Perth in two specials. In April of the following year, No 49 headed a special from Dundee to Glasgow in connection with an international football match. As on the great Carlisle excursion of 1899, McIntosh and MacDonald accompanied the party.

On the technical front, publicity was obtained by inviting Rous-Marten to sample the performance of the new engines, the outcome being a eulogy in *The Engineer* of 21 August 1903 under the title 'The New Caledonian giants at work'. By this time No 49 was at Polmadie and, with driver Ranochan at the regulator, was reserved for the Corridor, while No 50 was sent to Carlisle in charge of driver Will Todd, her regular roster being to take an early morning sleeping car express to Glasgow and return home with the 10am London express.

Rous-Marten's first run was made behind No 49 on the up Corridor, the gross load being 390 tons. Motherwell was passed at 35mph and speed did not fall below 30mph on the 1 in 102 through Wishaw South. There was a temporary fall to 27mph on the 1 in 100 near Carluke but by Craigenhill summit speed had risen to 31mph. After a brief spurt to 60mph beyond Cleghorn, the train

MCINTOSH DEVELOPMENTS 1901-4

was stopped by signals at Carstairs, the 29 miles having been run in 47min 38sec start to stop. The remaining 73.5 miles to Carlisle were covered in 81½min inclusive of two bad signal checks at Symington. These checks seem to have spurred the driver to greater effort. Assisted by about 1½ miles of 1 in 200 down gradient, speed rose from 20 to 57mph in 2½ miles and a vigorous assault was made on the remainder of the climb to Beattock Summit; on the two miles of 1 in 150 near Crawford, speed never fell below 47mph and the minimum on the final 2½ miles of 1 in 99 was 33mph, sustained for some distance. Despite the checks, the 23¼ miles from Carstairs to Beattock Summit were run in 34min 8sec. Rous-Marten remarked, 'Such a performance with so vast a load on such a gradient has never been equalled in my experience'.

Once over the summit, the 10 miles down the bank were covered in 9min 23sec and the remaining 39.7 miles to Carlisle in 37min 59sec, a maximum speed of 79mph being reached near Gretna Junction. Net time for the 102¼ miles from Glasgow Central was 124min and for the 73¼ miles from Carstairs 78½min. The engine had climbed well and shown herself to be as fast as the 4-4-0s. In Rous-Marten's words, 'I venture to think that such a performance is conclusive as to the merits of Mr McIntosh's new engine in respect alike of speed, of haulage power and of climbing capacity'.

On the evening of the same day, Rous-Marten returned to Glasgow on the down Corridor, the engine and load being the same as on the up run. The traditional brisk start was made, Rockcliffe (4.1 miles) being passed in 5min 57sec at 60mph. Speed rose to 64mph before Gretna and did not fall below 47mph on the first long rise at 1 in 200 through Kirkpatrick. On the slightly favourable (1 in 528) stretch north of Lockerbie, 69mph was attained, while on the four miles of 1 in 200 after Wamphray, 56mph seems to have been sustained, speed on the succeding 1 in 165 near Beattock still being as high as 53mph. Because the LNWR had handed the train over 3min late, the decision had been made to stop at Beattock for a banker if the lost time had not been regained by that point. In fact, the time taken for the 39.7 miles from Carlisle to Beattock was 44min 28sec, start to stop, only 28sec over the scheduled allowance. Then, with 0-4-4T No 443 at the rear, the 10 miles of 1 in 69-88 were run in 18min 32sec. The 23¼ miles from the summit to Strawfrank Jct, where the Edinburgh

MCINTOSH DEVELOPMENTS 1901-4

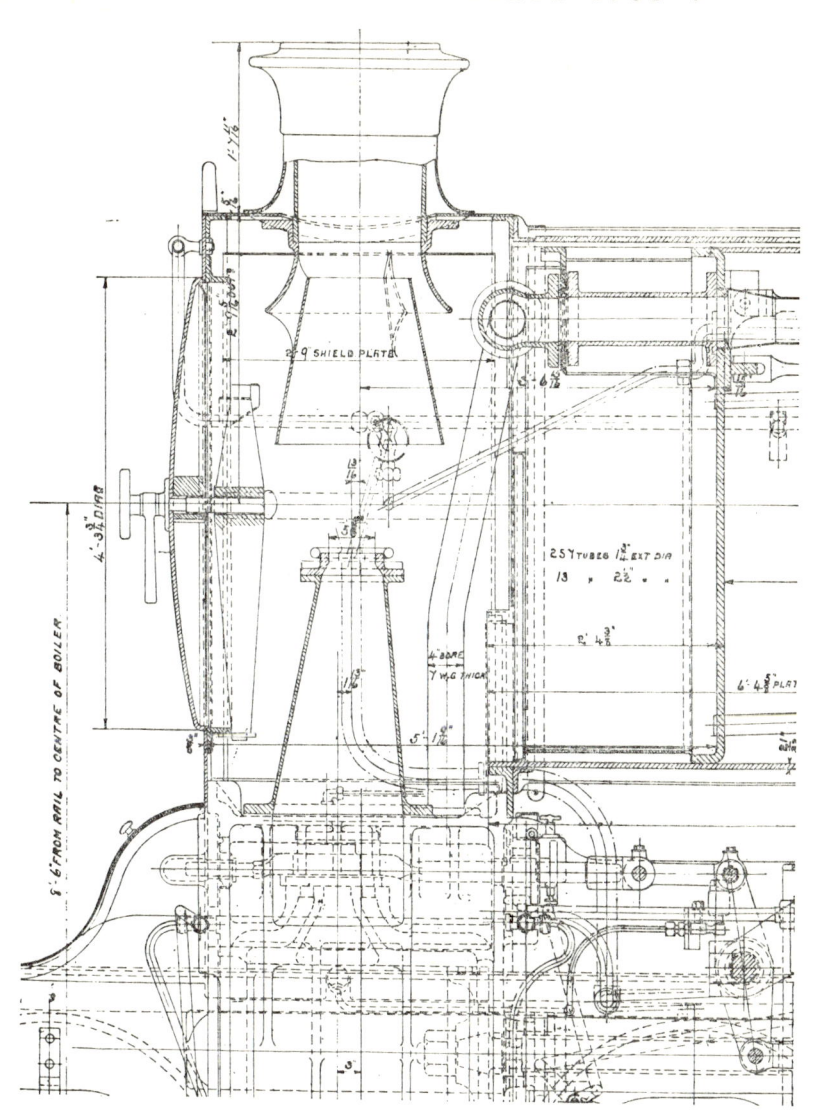

Fig 13   Smokebox arrangement of 49 class as altered

portion was detached, occupied only 21min 50sec, the maximum speed reached being 75mph. Net time for the 102½ miles to Glasgow was estimated at 109min, a remarkable achievement.

On his return to London, Rous-Marten travelled on the 10.45pm, hauled by No 50. This train was allowed 2hr 10min for the 102½ miles to Carlisle, inclusive of an eight-minute stop at Carstairs to take up the Edinburgh portion, the allowance of 80min for the 73½ miles from Carstairs requiring an average speed of 55mph. With 280 tons behind the tender, No 50 climbed the long bank to Craigenhill summit at minimum speeds of from 34 to 36mph and reached Carstairs in 40min 31sec start to stop. From there, with a load of 384 tons, the 15¾ miles to Abington took 20min 1sec and the 8.1 miles from there to Beattock Summit 11min 8sec, a steady 30mph being maintained up the final 1 in 99. The 49.7 miles from the summit to Carlisle were covered in 47min 5sec despite two signal checks. The net time for the 102½ miles from Glasgow was 110min, of which the 73½ miles from Carstairs occupied only 76½min net.

While the above performances were more than satisfactory, the new engines were not an unqualified success. In April 1903, a paragraph had appeared in *The Engineer* criticising the length of their boilers, the writer maintaining that £150 in cost and two tons in weight could have been saved on each engine had its barrel been 3ft shorter. Over the next six months, events proved that his misgivings were well founded for it emerged that the engines were rather shy steamers. The first modifications seem to have been initiated in October 1903 when the drawing office began work on a spare boiler. This differed from the original in having 230 2in tubes but the most significant change lay in the position of the smokebox tubeplate, this being recessed within the boiler-barrel by an additional 1ft 7in; length between tubeplates thus came down from 17ft 3in to 15ft 8in. New cylinder drawings, showing a reduction of 1in in the cylinder diameter, were prepared in May 1904 and the decision was made to modify the original boilers by recessing their smokebox tubeplate. These changes, together with a reduction of ⅛in in the diameter of the blastpipe orifice, were shown in a new general arrangement drawing dated 8 July 1904 and given the same number, 11600, as the original.

In September 1904, Richardson balanced slide-valves were substituted for the plain unbalanced valves of the original design.

MCINTOSH DEVELOPMENTS 1901-4

Slight modifications, involving a reduction in the clearance volume, were made to the cylinders in 1905.

THE 492 CLASS

In 1903 an eight-coupled mineral tank engine was designed, six engines being constructed:—

Order No	Date ordered	Engine Nos	Dates delivered	Cost per engine
Y71	5/03	492-7	11/03—1/04	£2250

The cylinders were of the standard 19in x 26in pattern with the valve-chests between them, and drove the second pair of wheels through standard 6ft 6in connecting rods. Since the leading and driving axles were 9in closer together than in the 600 class, and the intermediate axle was equidistant between the driving and the trailing axles, the wheel spacing appeared much less awkward than in the 600 class. The crank-axle was interchangeable with that of the 766, 812 and 900 classes. The driving wheels were flangeless and the coupling rods jointed at the driving and intermediate crank-pins, the intermediate axle having a side-play of ½in and the trailing axle one of 1⅙in. The springs of the leading and driving axles were not equalised but, in all other respects, the suspension was arranged as in the 600 class.

The boiler was a slightly modified version of that fitted to the McIntosh 'Jumbos'. The longitudinal girders on the roof of the inner firebox were linked to three, instead of two, angle-irons on the outer firebox casing, the rear two being connected by stout bars. This enabled working pressure to be raised from 150 to 175lb/sq in. Circulation in the water legs around the inner firebox was improved by a ½in increase in the width of the foundation ring while room was found in the tubeplates for an additional four tubes, bringing the total to 222. At 8ft 4½in above rail level, the boiler was pitched 1ft 1½in higher than that of the 'Jumbos' to provide adequate ashpan space, the grate passing over the intermediate axle. Even so, the bottom of the ashpan had to be curved up and over the axle and a plate inserted between them in order to protect the axleboxes and journals from ash and heat. On top of the firebox were two 4in lock-up safety-valves housed in a new form of casing, oblong in shape. The draughting arrangements were similar to those of the 600 class but the centre-line of the blastpipe orifice coincided with that of the chimney. Furness lub-

MCINTOSH DEVELOPMENTS 1901-4

ricators were fitted to the front of the smokebox, this being the last time they were used on a new design. Sanding was by gravity and reverse by unassisted lever. The cab was much longer than that of the 0–6–0Ts and was fitted with doors but the opportunity of extending it to the full width of the tanks was not taken. All six engines were Westinghouse-fitted and painted black, advantage being taken of the high pitch of the boiler to sandwich the main reservoir between the crank-axle and the lower surface of the barrel. Water capacity was 1450 gallons and coal 3 tons.

For many years No 496 was shedded at Dundee and was mainly employed banking goods trains on the Newtyle line. The remainder of the class were divided between Motherwell and Hamilton and could take slightly heavier trains than the 812 class, e.g., an extra three loaded 10-ton wagons or an additional five empty wagons on the Lesmahagow New Line. Most of their time was spent on short distance work. The Motherwell engines ran several return trips daily betwen the Lanarkshire Steel Works and Strathaven Jct, and between Milnwood Colliery and such centres as Mossend Bank and the Coltness and Clydesdale iron works. Those at Hamilton ran between Ross Jct and Coltness, and Ross Jct and Alton Heights.

THE 140 CLASS

In the autumn of 1903, work was begun on a larger-boilered edition of the 900 class. Although some 17 tons lighter in working order than the big 4–6–0s, the new 4–4–0s cost almost as much:—

Order No	Engine Nos	Dates delivered	Cost per engine
Y72	140-44	5/04–6/04	£3044
Y76	145-50	11/05–1/06	£2861

The cylinders, valves and motion were the same as those of the 900 class but the coupled wheelbase was lengthened by 3in. Laminated springs were hung below all four coupled axleboxes and the journals were enlarged:—

	Journals of driving axle		Journals of trailing axle	
Class	Dia, in	Length, in	Dia, in	Length, in
900	8½	7½	7¾*	9
140	9¼	8	7¾*	12

* Concave journals

The journals of the bogie axles also were enlarged and the driving axlebox guides were modified.

MCINTOSH DEVELOPMENTS 1901-4

The boiler was pitched 3in higher than in the 900 class and was 2¾in larger in diameter. The outer firebox was lengthened by 1in but an increase of 1in in the width of the foundation ring, though improving circulation in the water-legs, slightly reduced the grate area. The number of tubes was increased by seven, the lower 21 being of 2in diameter and the remaining 255 the standard 1¾in diameter. The blastpipe and chimney arrangements were based on those of the 900 class but with the proportions altered to suit the higher-pitched boiler. Above the firebox were four 4in 'lock-up' safety-valves enclosed in a cylindrical casing.

Reverse was by steam-assisted lever, the quadrant having nine notches on either side of mid-gear. Boiler feed was by two Gresham & Craven No 9 combination injectors. One engine of the second batch was fitted with compressed air sanding equipment but the other 10 engines got steam sanding apparatus. Pear-shaped displacement lubricators were attached to the smokebox front instead of the old Furness lubricators, while the lubricator for the Westinghouse pump was placed on the top of the cylinder.

The cab and splashers were in the 49 class style while the tender differed from that of the 4-6-0s only in its shallower (8½in) tank. There may have been some difference in the wells of the two batches for the locomotive register gave the water capacity of the first and second batches as 4450 and 4300 gallons respectively. All 11 engines were dual-fitted.

The new engines soon demonstrated their ability to pull hard and run fast but, with the introduction of the Cardean class 4-6-0s in 1906 and the coming of superheating in 1910, their reign on top-link duties was short. Outstanding runs recorded behind them were mostly made in 1905, when the class was distributed between Carlisle, Edinburgh and Perth:—

	Carlisle	Edinburgh	Perth
Engine Nos	142-3, 145-6, 148	140-41*	144, 149-50
	* On loan from Aberdeen		

No 147 is thought to have been at St Rollox shed.

Some remarkable work was done by these engines in their prime and it is fortunate that Rous-Marten was able to sample some of it, his article in *The Engineer* for 8 December 1905 giving details of three excellent runs, the first of which has become a classic in British locomotive history. The train was the 10.5am Edinburgh-

MCINTOSH DEVELOPMENTS 1901-4

London express, a very heavy formation extending from the buffer-stops at Princes Street to a van's length outside the platform end and estimated to have weighed 404 tons gross. The engine was No 140 with driver Stavert in charge and piloted to Cobbinshaw by rebuilt Conner 7ft 2-4-0 No 33. The remaining 82.2 miles to Carlisle were run unassisted in 87min 21sec start to stop, an average speed of 56.6mph. The most remarkable feature of the run was the all-out effort made up the final two miles of 1 in 99 to Beattock Summit.

> Here the speed at first dropped steadily as was certain to be the case with so vast a load on such a grade. In the end, it fell to exactly 36 mph, but at this point it kept steadily, quarter-mile after quarter-mile until the summit was reached, each quarter-mile being covered in exactly 25 seconds, as tested by two chronographs, whose respective starting and stopping buttons were pressed simultaneously as each quarter-mile post was passed. The engine never showed the least sign of flagging or slipping and was "going strong" as ever when the summit was breasted.

If Rous-Marten's words are taken literally, the equivalent drawbar horsepower required in maintaining 36mph would have been at least 1300 but if he happened to have been one-fifth of a second out on two successive quarter-miles, giving speeds of 36.3, 36 and 35.7mph over the last three quarter-miles, the horsepower developed would have been considerably less, though still high for an engine of the size. South of Beattock, some fast running was made, 76½mph being reached near Wamphray and 80½ on the descent to Gretna. Stavert, whom Rous-Marten once described as a 'veritable wizard', had previously had No 768, his usual run being the afternoon Edinburgh-London express, but when No 923 came out in 1907 he got her. Even then, however, he claimed that No 768 could be coaxed to handle as heavy a load as the 140 class engines and could run much faster.

Rous-Marten's second run was made with No 144 (driver 'Cuddy' Mitchell) on the up Grampian Corridor express when, with a load of 260 tons, the 32.5 miles from Forfar to Perth were run in 31min 1sec, start to stop, with a maximum speed of 83½mph. His third run was made with the Edinburgh portion of the 10am from Euston, loaded to 170 tons only and powered by No 141 (driver Watt). Despite bad weather and slippery rails, 60mph was reached before the end of the third mile and 75mph attained by

MCINTOSH DEVELOPMENTS 1901-4

the Solway Firth. The minimum on the two long 1 in 200 ascents that follow was 60mph and Lockerbie, 25.8 miles, was passed in 24min 55sec at 77½mph. The following 13.9 miles to Beattock were covered at an average speed of 67.2mph and a vigorous assault made on the bank. But at Auchencastle the train was stopped by a signal and the restart had to be made on the 1 in 75 gradient in drizzling rain. Remarkably, speed was quickly worked up to 36mph and although there was some slipping on the curves at Greskine, causing a fall to 33mph, 36mph was regained and maintained unbroken right to the summit. The 10-mile bank was climbed in 19min 38sec, inclusive of the dead stand, which occupied 2min 43sec, the equivalent drawbar horsepower involved being around 900.

THE KEYSTONE OF THE ARCH

Between January 1895 and January 1899, the board had authorised the construction of 192 engines to capital account. The sum provided—£442,466—was very high for a Scottish railway and it was largely due to the board's willingness to invest money on this scale that the company had reached its pre-eminent position. The last of these engines was built in the half-year ending 31 January 1901 and from then until 1906 no further locomotives were built to capital account. While in the long term this was to be regretted, the company nevertheless maintained a good record of locomotive construction out of its revenue account. During the ten years ending 31 July 1905, 177 new engines were constructed out of revenue at a cost of £339,781 and a further £100,000 was spent on rebuilding, this being additional to the money spent on repairs. These sums were in sharp contrast to those spent by the NBR during the same period:—

	Caledonian	NBR
Capital	£442,466	£329,439
Revenue	£439,781	£196,506

The average cost of a new engine was the same in both companies, namely £2304. As far as expenditure on repairs was concerned, the Caledonian spent a much higher proportion of the total on materials and a much lower proportion on wages. In a letter to McIntosh, dated 1 November 1905, Dr John Inglis, a director of the NBR and a member of its locomotive committee, wrote:—

115

MCINTOSH DEVELOPMENTS 1901-4

Some time ago, I was chatting with Sir James Thompson in the CR office when he said that the differences between the CR and the NBR costs of renewals and repairs alarmed him. I replied that they alarmed me also but for a different reason—He thought the Caley costs too high whereas I was of the opinion that the NB were not high enough.

McIntosh replied two days later:—

> I am also perfectly at one with you in regard to the futility of repairing old, light, antiquated engines and I am pressing our people strongly as to the desirability of breaking them all up and spending the money in building new modern engines, which would otherwise be spent in repairing old light locomotives, which even after repair are not fit to handle heavy plant and trains satisfactorily, nor with the necessary facility that is required nowadays.

It would seem, however, that certain of the directors were rather parsimonious in their attitude to investment. On 31 October 1905, McIntosh informed Inglis:—

> I have always a difficulty in getting our directors to understand why the CR accounts do not appear to come out so well as the NB. You will readily see that this is in consequence of the new stock we build out of revenue, which causes our expenditure to be higher than the NB.

But the happy state of affairs on the Caledonian had then already ended, as McIntosh tacitly confessed to Inglis on 27 October.

> You are perfectly right in your statement that 1/30 of the total number of engines should be renewed every year out of revenue and we have hitherto always acted on this principle, with the exception of the last year or two when we have been curtailing our expenditure and, last year for instance, only 20 new engines were built out of revenue and 10 rebuilt.

This cut-back in expenditure rang the death-knell of an effective big engine policy; after looking to the renewal of its numerous local passenger, goods and mineral engines, there was no money left to pursue a bold big engine policy.

CHAPTER 8

CLIMAX AND DECLINE 1904-9

CARDEAN

Tom MacDonald retired in the autumn of 1908 and was succeeded by the assistant works manager, Irvine Kempt, Junior, at a salary of £300, his job in turn passing to John G. Barr at £200. It was the beginning of the end for the old guard and the emergence of a new generation of men.

On 14 November 1905, the board approved the construction of 20 10-wheeled engines out of capital, in addition to the locomotives allowed for in the half-yearly building programme:—

> *New engines to Capital Account.*
> Submitted the General Manager's memo of 6th inst, recommending that the Directors should authorise the construction of 20 additional engines at a cost of £55,000, viz.,
> 10 six-coupled bogie express goods engines intended for express goods trains between Glasgow and Carlisle and between Glasgow and Aberdeen.
> 5 six-coupled bogie passenger engines to be used for the Aberdeen and Glasgow Corridor express trains, which are now heavy and booked to run at high speeds.
> 5 bogie "Atlantic" type of engines to work the Edinburgh and Carlisle express trains and the down tourist and postal trains between Carlisle, Perth and Aberdeen.
> 10 engines to be built during the half-year ending 31st July 1906 and the remainder during the half-year ending 31st January 1907.
> Approved.

McIntosh abandoned the idea of an Atlantic and built instead an improved larger-boilered version of the 49 class:—

Order No	Engine Nos	Dates delivered	Cost per engine
Y80	903-07	5/06 – 7/06	£3500

CLIMAX AND DECLINE

The cylinders were reduced to 20in in diameter and, to allow for longer crank-axle journals, were brought 1¼in closer together than those of the 49 class. The steamchests were above the cylinders and housed a pair of Richardson balanced slide-valves. The total wheelbase remained as before but the driving wheels were

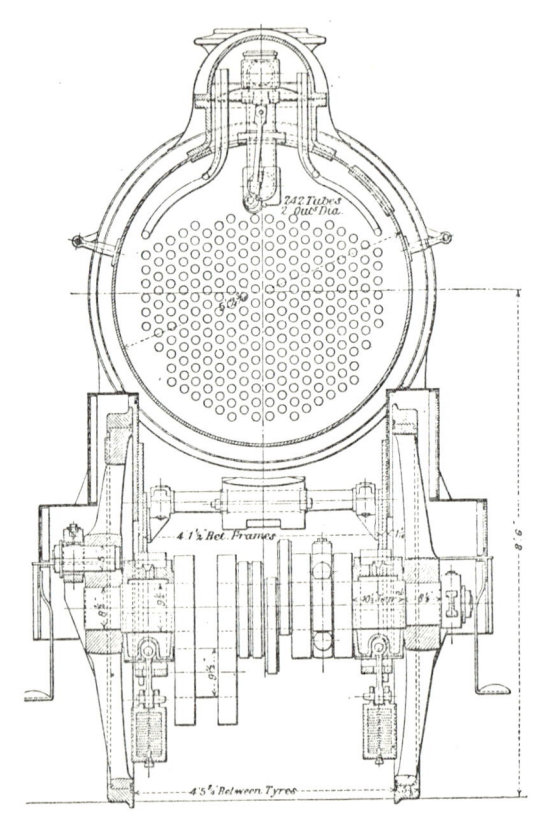

Fig 14   903 class: section through crank-axle showing dished wheels

moved 4in closer to the intermediate coupled wheels, this allowing a little more room for access to the valve gear. The connecting rods were lengthened to 7ft but the eccentric-rods remained at 4ft 6in. The solid-forged crank-axle, with its elliptical webs, was replaced by one of built-up design, with parallel-sided webs, manufactured by a private contractor. To provide space for larger crank-axle journals than before, the leading coupled wheels were dished outwards in the centre, the inner surface of the wheel-seats thus being several inches further from the frames than that of the tyres. The journals were lengthened by 1in. The intermediate and trailing coupled wheels were of normal design but their journals were concave and longer than those of the 49 class:—

Class	49	903
Driving journals: diameter	9¼in	9¼in
length	9½in	10½in
Intermediate and trailing journals: diameter	8½in	8-9½in
length	9in	12in

The driving and intermediate axles were allowed ⅛in side-play and the trailing axle ½in, the two sections of the coupling-rod being united by a knuckle joint. The axlebox guides were of a new pattern while the balance weight on the weighshaft consisted of a heavy cylinder suspended between backward extensions of the two lifting arms. In other respects, the motion was similar to that of the 49 class though several of its components were stronger, the big-ends being ¼in wider and the intermediate and trailing axles ¼in greater in diameter. These alterations probably permitted the engines to work harder for longer periods than their predecessors and with less risk of running hot. But in one important respect the new engines were at a disadvantage, at least in the long term: in order to keep their weight down to the 73-ton mark, the thickness of the frames was restricted to 1 1/16 in Another questionable step was the reversion to gravity sanding.

Steaming capacity was improved in several ways. With an outside diameter of 5ft 3½in, the rear ring of the boiler was 3½in greater in diameter than that of the 49 class. The barrel was lengthened by 3⅜in but, with the smokebox tubeplate recessed 1ft 4⅜in within the front ring, length between tubeplates was 16ft 8in, 1ft longer than that of the 49 class in their modified form. Heating surface was increased by the use of 242 2in tubes and the proportions of the firebox altered slightly:—

Class	49	903
Overall length	8ft 6in	8ft 6in
Depth below boiler centre-line, front	5ft 3in	5ft 0in
rear	3ft 9in	3ft 9in
Height above boiler centre-line: inner firebox	1ft 1in	1ft 3¼in
outer firebox	2ft 6in	2ft 7¼in
Width of foundation ring	2¾in	3in

CLIMAX AND DECLINE

Circulation over the inner firebox crown was improved by the substitution of four rows of flexible stays and 17 rows of plain stays for longitudinal girders. The brick arch was shorter and much more steeply inclined. The ashpan was redesigned, the rear being extended down behind the trailing coupled axle to form a

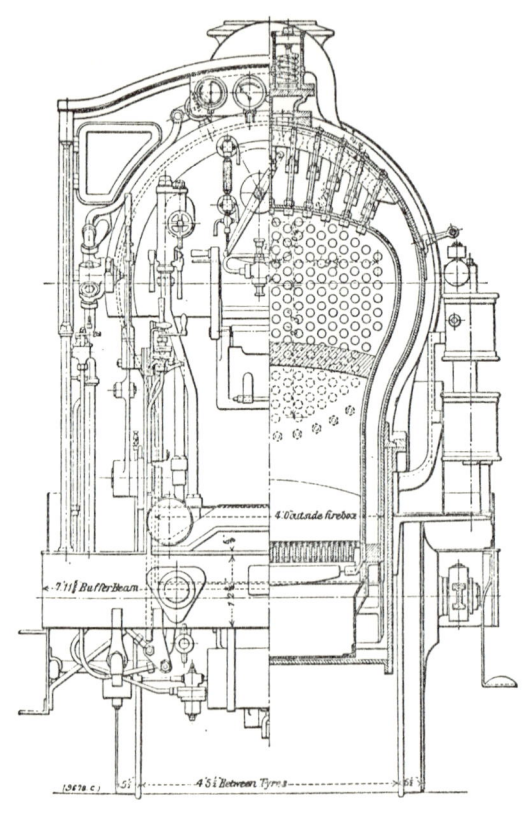

Fig 15 903 class: section through firebox showing direct crown stays

sort of hopper, at the bottom of which a damper was placed. The smokebox arrangement was similar to that of the 49 class but the diameter of the blastpipe orifice was reduced to $5\frac{1}{2}$in. Tests showed that the boiler could produce more steam than that of the 49 class, the improved circulation around the inner firebox being regarded as the main contributory factor.

CLIMAX AND DECLINE

A number of modifications of lesser significance were also made. The boiler-barrel was composed of four telescopic rings and the safety-valves were housed in two parallel oblong casings. The heavy cast-iron dragbox of the 49 class was replaced with one of steel plate and the brake and cylinder cock riggings were altered. Levers were provided for the gravity sanding and the additional damper. In the 49 class, the vacuum ejector had obstructed access to the blower handle so, in the new engines, the handle, which was of a new stirrup-shaped pattern, was moved well back into the cab. The reversing quadrant was faced with a brass plate and had 12 notches on either side of mid-gear. The reversing lever was powered by a 6in diameter steam cylinder in place of the standard 4½in diameter pattern. Front-end lubrication was arranged as in the 49 class but instead of Furness lubricators two pear-shaped displacement lubricators were provided.

The 903 class was introduced without publicity but the class became something of a legend. No 903 was named *Cardean* after the estate of the deputy chairman, Edward Cox, the name appearing in gilt letters on the leading splashers and the coat of arms on the middle ones. She was then sent to Polmadie and, with James Currie at the regulator, replaced No 49 on the 2pm up and 8.13pm down Corridor, No 49 being put on the 4.30pm up Liverpool and the corresponding down working from Carlisle. No 905 went to Dalry Road in charge of the famous Will Stavert, while Nos 904 and 906 were sent to Perth and No 907 to Carlisle. Dalry Rd soon lost No 905 to Perth and No 906 was transferred to Carlisle about the same time. With the arrival of No 906, No 50 was transferred to St Rollox. Just when this move was made is not clear but she was seen on the Carlisle road as late as 1909.

Following transfer, No 905 worked in a link with No 904, the turns being the 11.40am Perth-Aberdeen, returning with the 5.30pm Aberdeen-Perth, and the 3.20pm Perth-Aberdeen, returning with the 7.50pm London sleeper. After a short time this train was handed over to an Aberdeen engine and the 4–6–0 returned with the 8pm fish. One of the Carlisle engines ran the 4.22am Postal to Glasgow and returned with the 10am Glasgow Central-London express, while the other took the 5.15am sleeper from Carlisle to Glasgow and brought back the 10.10am Glasgow-Liverpool express. This duty was later allocated to a 4–4–0, the 4–6–0 then taking the 3.55pm Carlisle to Glasgow and the 10pm

CLIMAX AND DECLINE

London sleeper from Glasgow Central. From St Rollox No 50 worked the 2pm Buchanan Street-Perth, made a return trip to Dundee, returned south with the 7.45pm Perth-Glasgow Central and finally ran the empty stock back to Buchanan Street via Rutherglen and Parkhead.

The 4–6–0s had an ample margin of power for most duties and were seldom unduly pressed. This applied particularly on the Aberdeen road where, with loads rarely exceeding 240 tons, the 11.40am Perth-Aberdeen was allowed 2hr 5min inclusive of four stops, and the return 5.30pm 2hr, again with four stops. The up Corridor ran non-stop to Carlisle in 2¼hr while the down train was allowed 2hr 3min inclusive of a two-minute stop at Beattock and another of the same duration at Strawfrank Junction for the detachment of the Edinburgh portion. Of the Carlisle turns, the 4.22am down was allowed 2hr 13min with three stops and the 10am up 2hr 12min with one stop. The 5.45am down was booked in 2hr 5min with one stop and the 10.10am Glasgow-Liverpool in 2½hr inclusive of four stops. On these duties, loads were between 305 and 405 tons gross. There was seldom opportunity to extend the engines and, as Table 6 shows, there was sometimes little to distinguish the work of the 4–6–0s from that of the 140 and 900 classes.

TABLE 6
1907: 2.50pm SYMINGTON–CARLISLE

Run No	1	2	3	4	5
Engine No	141	905	899	899	896
Load	=21	=15½	=16½	=14	=10½
Weather	Wet	Fine	Wet	Fine	Fine
		Strong side wind			

	Miles	Time, min sec				
Symington	0	0	0	0	0	0
Beattock Summit	17.1	25 10	23 41	24 25	22 54	21 30
Beattock	27.2	35 05	33 33	34 06	32 36	32 14
Lockerbie	41.1	47 50	45 55	47 37	45 42	46 50
Kirtlebridge	50.2	57 05	55 53	57 28	54 47	56 34
Gretna	58.3	64 03	63 02	66 05	62 42	64 43
Carlisle	66.9	74 02	73 18	75 05	72 15	74 40

Delays: Run No 2 pws Ecclefechan 1min
3 pws Kirtlebridge 1min
4 pws Gretna ½min

Page 123 (upper) 908 class 4–6–0 No 917 on Gourock train at Glasgow Central; *(lower)* 179 class 4–6–0 No 184 at Perth, June 1921

Page 124 (*upper*) 300 class 0–6–0 No 313; (*lower*) 60 class 4–6–0 No 63

CLIMAX AND DECLINE

Judging from the performance of *Cardean* on the LNWR in July 1909, the 903 class was capable of magnificent work. In the LNWR trials initiated by C. J. Bowen-Cooke, *Cardean* arrived at Crewe with the up Corridor on 15 June and ran between Crewe and Carlisle until 10 July. The highlight on 6 July has become one of the classic performances in British locomotive history. The train was the 12.58pm from Carlisle. Only 24min, start to stop, were allowed for the 17.9 miles to Penrith and, after a five-minute stop, 18min were allowed for the 13.6 miles to Shap Summit—timings sharper than any operating during the inter-war period save that of the *Coronation Scot*. With a tare load of 367 tons, *Cardean* was set a formidable task. By developing a sustained output of about 1200edhp between Carlisle and Plumpton and between 1400 and 1500edhp for no less than 16 consecutive minutes south of Penrith (O. S. Nock's calculations), *Cardean* kept to this most demanding schedule. Her performance is all the more praiseworthy when it is realised that she had arrived at Carlisle after a spirited run from Crewe only 10 minutes before her departure for the South. With a load of 301 tons tare, she had run the 31.4 miles between Carnforth and Shap Summit in $42\frac{3}{4}$min with speeds of 60mph at Burton and Holme, 37 at Oxenholme, 33 at Grayrigg, 63 at Tebay and 31 at Shap Summit, drawbar pull at these five points being $1\frac{3}{4}$, $2\frac{5}{8}$, $3\frac{5}{8}$, $1\frac{7}{8}$ and $5\frac{3}{8}$ tons respectively.

Not long before the trials, *Cardean* had been involved in a spectacular derailment. She had entered St Rollox works for major overhaul on 26 August 1908 and emerged again on 21 October. Although her wheels had been inspected, they had not been removed from the axles, the wheel-seats therefore missing examination. Her built-up crank axle had been supplied by John Spencer & Sons Ltd with a guarantee that it would be replaced if it failed before it had run 200,000 miles. On 2 April 1909, she had just passed Crawford on the down Corridor when her crank-axle—which had amassed a mileage of 145,389—broke between the left wheel-seat and the journal.

Following the accident, intensive tests were carried out on what remained of the broken crank-axle. The report concluded: 'The axle had been made of inferior metal which had been left in an overheated condition and suffered from fatigue in use'. In his official report on the accident Colonel Yorke criticised the design of the failed portion of the axle and, at its meeting of 9 July 1909, the locomotive and stores committee 'resolved that all passenger engine axles of the kind objected to by Colonel Yorke should be altered forthwith'. New drawings for the 903 class crank-axle and driving wheels were prepared in August 1909, the diameter of the wheel seat being increased to match that of the journal.

	Old in	New in
Wheel seat: length	7½	8
diameter	9	9¾
Journal: diameter	9½	9¾
Wheel boss: radius	8¾	10¾

THE 918 AND 908 CLASSES

A further 15 4–6–0s of the special capital programme were turned out between July 1906 and January 1907.

Order No	Engine Nos	Dates delivered	Cost per engine
Y79	918-22	7/06–9/06	£2871*
Y81	908-17	10/06–1/07	£3031

* Westinghouse fitted. £3100 for dual fitted engines

This was a reversal of the original decision to build 10 express goods engines and five passenger engines. The 918s were basically 55 class engines with the trailing overhang and outer firebox lengthened by 6in and with a much-shortened version of the 903 class boiler-barrel, length between tubeplates being only 13ft 6¼in as in Nos 51-4. The frames, front-end, motion, wheels, steam reversing gear, sanding and brakes were laid out as in the 55 class but a knuckle joint was provided in the coupling rod to allow the trailing axle a side-play of ½in. The frames were 1⅛in thick and the driving journals increased in diameter:—

Class	Diameter	Length
55	8½in	7½in
918	9¼in	7½in

The crank-axle was of the built-up type with parallel-sided webs and the intermediate and trailing journals were of concave form. Front-end lubrication was arranged as in the 55 class but with the new displacement lubricators in place of the Furness type.

The boiler-barrel was composed of four telescopic rings and contained 242 steel tubes of 2in diameter. Working pressure was

CLIMAX AND DECLINE

175lb/sq in, the safety-valves being of the size and type used on the 903 class. The firebox was better proportioned than that of the 55 class:—

Class	55	918
Overall length	6ft 5in	6ft 11in
Depth below boiler centre-line: front	5ft 0in	5ft 0in
rear	4ft 3in	4ft 3in
Height above boiler centre-line: inner firebox	1ft 0½in	1ft 3¼in
outer firebox	2ft 4⅜in	2ft 7¼in
Heating surface, sq ft	105	128
Width of foundation ring	2½in	3in
Grate area, sq ft	20.63	21

Circulation was helped by the wider water legs and by the substitution of direct stays for the longitudinal girders on the roof of the inner firebox. The design of the ashpan followed that of the 903 class, and a rear damper was provided. Although vacuum ejectors and dual brake handles were fitted to Nos 918-9 only, all five engines got the 903 class type of blower handle. The cab windows were of a new form, the change from the 903 class pattern being necessary because of a 4in reduction in the width of the cab. The tenders were of the 3570-gallon pattern. Although intended solely for goods traffic, the engines were painted blue.

Nos 918-20 and 922 went to St Rollox and No 921 to Dalry Rd. From St Rollox, they did double trip working to Carlisle, one diagram covering the 12.2am and the 9.15pm from Buchanan Street (Goods). Other trains they handled were the 12.40pm and 9.35pm from St Rollox to Aberdeen. They remained on long-distance goods trains until July 1914 when Nos 918 and 920-1 were sent to Oban to help with the heavy holiday traffic. Coal consumption is said to have been very heavy; the high free gas area/grate area ratio, together with the relatively short tubes, probably resulted in a high rate of combustion and, with the soft Scottish coal, it is likely that fuel passed out of the chimney unburned.

The 908 class was a larger-wheeled version of the 918 class. The frames were lengthened by 1ft 5in to accommodate 5ft 9in coupled wheels and length between tubeplates was increased to 14ft 11¹⁄₁₆in. The trailing axle was closer to the rear of the firebox than in the two preceding classes; the reduction in space between

CLIMAX AND DECLINE

the axle and the injector pipes precluded the use of the 918 class ashpan and one of 55 class pattern was provided, with front damper only. The brake-rigging was based on that of the 903 class. The steam reverser was bolted to the inside of the frame, in front of the leading splasher, and sandpipes fitted for forward running only, the sandboxes being located below the running-plate. Steam heating equipment was fitted and vacuum ejectors, combined brake handles to Nos 908-9, 912 and 916-7. The cabs and tenders were similar to those of the 918 class, the tenders carrying gilt scrolls on either side of the coat of arms.

No 909 was named after newly-appointed chairman, Sir James King, and No 911 after Barochan, the estate of Sir Charles Bine Renshaw, MP. Nos 908 and 913 were allocated to Perth for working the principal expresses between Glasgow, Perth and Dundee. When new, No 908 was the regular engine on the 10am Grampian Corridor Express between Buchanan Street and Perth but after several years was succeeded by No 913. She continued as regular engine for a year or two until the principle of 'same engine daily' was relaxed. The engine worked up to Glasgow with the 7.15am from Perth and, after its return, completed the day's work with a return trip to Dundee. Despite their small fireboxes, which called for some very skilful firing, the Perth engines did some competent work. In 1913 No 913 on the 10am from Glasgow, with a load of 285 tons full, and without banking assistance, made a very good job of the gruelling climb out of Buchanan Street. Good work was also done on the other uphill stretches and in accelerating away from checks. On the level stretch before Dunblane bank, speed rose from 20 to 52mph in two miles and on the bank, 5½ miles of 1 in 100-88-134, the minimum was 28mph. With speed not exceeding 65mph at Auchterarder and despite a further permanent way slowing, the train was comfortably on time on the approach to Perth. In skilled hands the class did all that was asked of it.

Nos 909, 911-2 and 914 worked between Glasgow Central and the Clyde coast, *Sir James King* and *Barochan* being shedded at Greenock. A distinct improvement on the 'Coast Bogies' and 0–6–0s, they took over the crack Gourock workings such as the 4.5pm from Glasgow Central. It was a time of intense competition for the rail-steamer traffic to the Clyde resorts and, in terms of advertising value alone, the engines probably justified their construction.

CLIMAX AND DECLINE

QUIESCENCE

Some four years elapsed before another new design left the works. Additions were made to existing classes and thought given to the design of boilers, cylinders, motion, etc., but, with one or two exceptions, little of long-term significance was achieved.

From 1904 onwards, the stock of 782 class 0–6–0Ts was gradually expanded, four more batches being turned out prior to 1910:—

Order No	Engine Nos	Dates delivered	Cost per engine
Y73	631-35	10/04—11/04	£1595
	636-40	11/04—12/04	£1655
Y74	641-50	3/05—5/05	£1655
Y82	128-9, 166, 324, 472, 668, 425, 434, 501, 630	4/07—6/07	£1416
Y83	417, 426, 629, 667, 669	9/07—10/07	£1410

These engines differed slightly from the early batches. The firebox crown was strengthened by the provision of an additional link between the outer firebox casing and the longitudinal girders on the roof of the inner firebox and this allowed working pressure to be increased from 150 to 160lb/sq in. Circulation in the waterlegs around the inner firebox was improved by a half-inch increase in the width of the foundation ring. The draughting arrangements were also changed; the blastpipe of batches Y73-4 was shorter than that of the earlier engines and the diameter of its orifice was increased from 4¾in to 5in, while a hood and petticoat were added. With batches Y82-3 a return was made to a 4¾in diameter orifice and the hood and petticoat were abandoned. Instead of Furness lubricators on the smokebox front a pair of the new displacement lubricators were provided while several changes were made in the layout of the cab fittings; the sight-feed lubricator was transferred from the left to the right-hand side, and the steam brake cock, which had previously been at the extreme right of the faceplate, was moved to the left of the adjacent injector. Minor external modifications included the substitution of single front footsteps for the double ones previously used. Nos 636-50 were Westinghouse-fitted but, like their steam-braked sisters, were painted black. Of the 35 engines, just over one-third went to Polmadie. Dawsholm and Grangemouth got about five each, while the remainder were distributed in ones, twos and threes to Motherwell, Hamilton, Dalry Rd, Carstairs and St Rollox.

CLIMAX AND DECLINE

Roughly contemporary with the 0-6-0Ts were four batches of standard 439 class 0-4-4Ts:—

Order No	Engine Nos	Dates delivered	Cost per engine
Y77	151, 473, 655	3/06	£1864
Y78	125, 384, 660	4/06	£1864
Y78	112, 424, 464-6, 666	12/06—2/07	£1910
Y84	158, 419, 422, 429, 470	11/07	£1627
Y90	126-7, 420-1, 427-8, 456, 467-9	10/09—12/09	£1445

These engines had the same boilers and lubricators as the contemporary 0-6-0Ts. The smokebox arrangements were probably similar to those described for the latter class but engines of batch Y90 got the new form of hood and a very tall blastpipe. Approximately eight of the class were allocated to Forfar and its sub-sheds and six or seven to Polmadie. Perth, Dundee, Stirling, Motherwell, Hamilton, Dalry Rd and Beattock each got one or two.

After an interval of six years, another two 0-4-0STs were built:—

Order No	Engine Nos	Dates delivered	Cost per engine
Y88	431, 463	12/08	£774

These were the last of the class.

In 1907-8, a further batch of the 140 class was turned out:—

Order No	Engine Nos	Dates delivered	Cost per engine
Y85	923-7	12/07—2/08	£2874

These engines differed from Nos 140-50 in having built-up crank-axles and gravity sanding. The safety-valves were enclosed in twin oblong casings, as in the 903 class. All were dual-fitted. Nos 923-4 were allocated to Edinburgh and Nos 925-7 to Aberdeen, later Carlisle.

Following the 4-4-0s came three batches of 0-6-0s, these being a modernised version of the 812 class:—

Order No	Engine Nos	Dates delivered	Cost per engine
Y86	658-9, 662-4	6/08—9/08	£2088 with vacuum ejectors
Y87	652-4, 656-7, 665	3/08—6/08	£2055 with steam brake only
Y89	325-8, 423, 460	6/09-7/09	£1682

130

CLIMAX AND DECLINE

Externally, the main changes lay in the shape of the cab and frames. The cab was modelled on that of the 900 class while the frames were heightened in the region of the driving axleboxes to project above the splashers. The cylinders, axleboxes, cylinder cocks and sanding gear were slightly modified, while laminated springs replaced the coil springs below the driving axleboxes. The blastpipe was of the type used on the 812 class from 1905 onwards (see below). The engines were steam-braked and painted black. Nos 652-3, 657-8 and 665 also had vacuum ejectors. Nos 325 and 328 went to Polmadie, No 326 to Aberdeen and No 661 to Dalry Rd. Most of the remainder were shedded at Carlisle, their number including Nos 327, 652, 654, 656-9 and 662-4, and worked on long-distance goods.

In 1906 a new boiler was designed for the 721 class, the original dimensions being retained but with 275 1¾in tubes and altered staying. No 732 was the only engine of the class officially rebuilt (1908) but some of the others probably acquired the 1906 boiler. New boilers were designed for the Drummond 0-4-4Ts in February 1908 and were fitted as the old ones wore out though, with the exception of No 175, dealt with in 1901, and Nos 172 and 174,

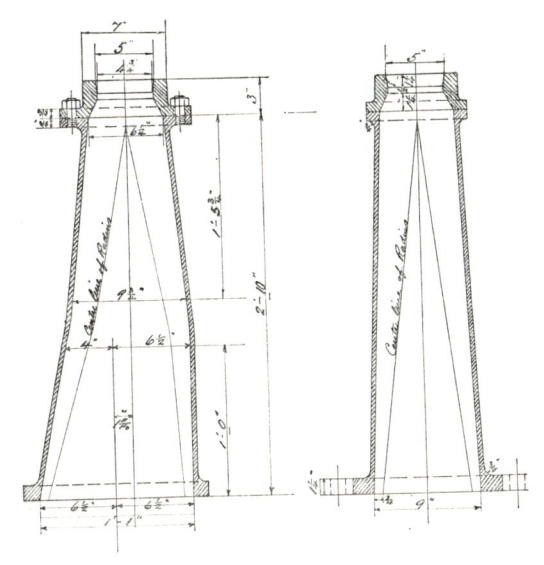

Fig 16   104 class blastpipe: 1904

similarly treated in 1904, none of the class was officially recorded as rebuilt prior to the grouping. The 1908 design differed from the original in having a flat-topped inner firebox stayed by longitudinal girders and only 134 tubes, but retained the safety-valves on the dome.

As with the boilers, the cylinders of the earlier engines were updated during the early 1900s, the changes being relatively minor:—

Type of engine	Date cylinders redesigned
18in x 26in and 18¼in x 26in 4–4–0s	11/04
14in x 20in saddle tanks	11/07
18in x 26in 0–4–4Ts	4/09
18in x 26in 0–6–0Ts	5/09

Blastpipe design was changed also, the vortex pattern being replaced by several varieties of plain pipe. Thus:—

Class	Plain blastpipe designed
104	6/04
812	3/05
171	9/07
721	11/07
'Coast Bogie'	12/07

An important feature of blastpipe design during this period was the tapered orifice, shown in the accompanying drawing. Diameter apart, the same dimensions were adopted for the blastpipe caps of engines as diverse as the 908 class 4–6–0s and the outside-cylinder saddle-tanks. Another noteworthy feature was that the centre-line of the blastpipe was placed slightly ahead of the centre-line of the chimney. This does not seem to have had the adverse effect on steaming that theory and other experience would indicate.

During the early years of the century, the company paid out some £2000 per annum in compensation for damage caused by sparks from locomotives. A spark-arrester was designed in 1908 and fitted to the 652 class 0–6–0s from No 664 onwards. It was not until the following summer, however, that the apparatus was applied generally. By then it had been thoroughly tested on engine No 147 and patented under the names of J. F. McIntosh and W. R. Preston, a director of J. Stone & Co of Deptford (Patent No 9994, deposited 29 April 1909). The principal components of the apparatus were a V-shaped plate, attached to the rear of the blastpipe and forming a shield between it and the tubeplate, a 3in

CLIMAX AND DECLINE

diameter suction pipe reaching from the bottom of the smokebox to a point within the chimney liner, and a series of louvres on the inside of the smokebox door.

The trials were held in January 1909 and were intended to show what happened inside the smokebox in running conditions. A 4in square was cut in the side of the smokebox and a thick glass plate inserted. A wooden shelter was built alongside the smokebox and

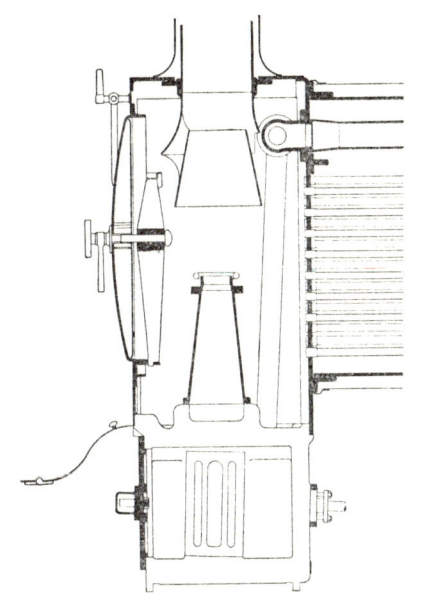

Fig 17 140 class: draughting arrangement as originally designed

the locomotive, complete with observer, sent out on the 6.10 pm from Buchanan Street to Kinbuck. From that point No 147 ran forward to Perth light. It then took the 9.40pm Perth to Coatbridge, and finally worked back light to St Rollox where the staff were waiting to examine the data collected.

With the standard blastpipe and chimney arrangement and without the spark arrester in position, it was found that red hot cinders flew out of the tubes at high velocity and continued in a straight line through the exhaust jet to impinge upon the smokebox door, sparks passing as little as ¼in above the blastpipe orifice without being deflected. Small cinders from the smokebox door were then drawn into the chimney but large pieces fell to the bottom of the smokebox and formed a drift against the door. At first there was little emission of sparks but, as the drift built up, a fountain of red embers erupted from the chimney. After 13 miles of climbing, the drift was half-way up the smokebox, with the highest part against the door. By then, cinders from the tubes were racing up the face of the drift and passing straight out of

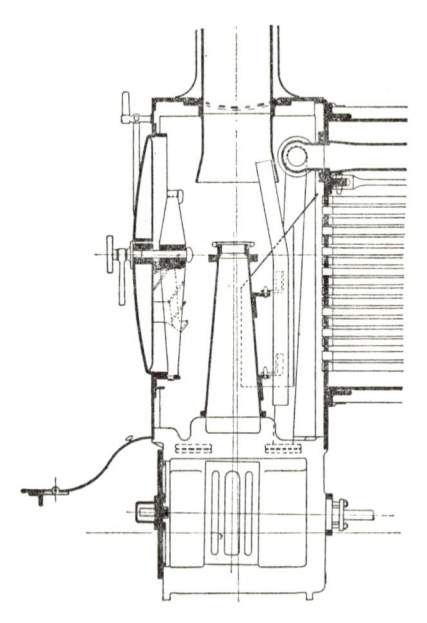

Fig 18 140 class: spark arrestor and altered draughting arrangement

the chimney without having had time to cool. With the louvres and V-plate in position, the cinder bank built up in the angle of the plate, the louvres preventing cinders from piling up against the door, but when the suction pipe was added, the 'cool' cinders from the bank were drawn into the chimney. The drift no longer provided a rapid, direct route for the sparks. Red hot embers ejected from the tubes progressed backwards to the foot of the pipe in a series of jerks and, within seconds of entering the smokebox, had cooled to invisibility.

Since thought was also being given to a complete re-draughting of the company's locomotives, No 147 was run with altered

CLIMAX AND DECLINE

as well as with the original draughting arrangements. The shape of the hood was changed, the upper two-thirds now being parallel and the lower portion conical. The petticoat was discarded and the blastpipe increased in height, its orifice now being just above the boiler centre-line. By this time, the vortex blastpipe was a thing of the past and the only engines for which new blastpipes had not been designed were the 4-6-0s, the 0-8-0s, the condensing engines and certain of the 'Jumbos'. During 1909, and beginning with the 49 and 903 classes, a start was made in fitting spark arrestors and the new type of hood to engines going through the works, the only classes not so treated being the outside-cylinder saddle-tanks. At the same time, taller blastpipes were designed for the 55, 171, 439, 492 and 782 classes and for all of the 4-4-0s, the remaining classes combining the new form of chimney with contemporary patterns of blastpipe.

CHAPTER 9

THE COMING OF SUPERHEATING

REFLATION

In March, 1909 a committee was appointed to inquire into the administration, costing and accountancy at St Rollox. Its report showed that the capital value of the company's rolling stock had not varied much during the past five years and that the average sum spent per annum on the renewal, ie replacement, of its locomotives was only £30,484, or 1.6 per cent of their capital value. If the number of locomotives was to be held constant and their rate of replacement not increased, their average age on replacement would be 70 years, approximately double the lifespan which could reasonably be expected. The report concluded:—

> It is obvious that the above-mentioned rate of Renewals is not sufficient, and that the Capital invested in rolling stock is not being adequately protected. It is necessary to adopt some standard of Renewals, and it is suggested that this be $2\frac{1}{2}\%$ per annum of the Capital value of engines and carriages. This represents an assumed life for each unit of 40 years.
> As regards engines, Renewals equal to $2\frac{1}{2}\%$ on the Capital value of £2,135,945 would cost £53,399 per annum, say £27,000 in each half year. This represents an increase of £11,500 per half year over the average of the past 5 years. Considerable economies in Repairs may be expected as an immediate result of such increased Renewals. Also, Revenue must benefit by the increased tractive power and efficiency of the newer engines. It is probable that the charge to Revenue for Repairs and Renewals of Engines would be increased to an amount considerably less than £11,500 per half year.

The board resolved that a renewal charge at the rate of $2\frac{1}{2}$ per

THE COMING OF SUPERHEATING

cent be the ultimate aim and that, as a step in this direction, the renewals for 1909 be arranged on the basis of a charge to revenue of 2 per cent. The 2½ per cent renewal charge was not achieved for several years. In a memorandum of 4 September 1911, the general manager reminded the board:—

> The matter of the renewal of the Company's rolling stock is one that has been repeatedly considered by the Chairman and Directors. While provision for the renewal of engines and carriages is meantime fixed on a basis of 2¼% per annum of the Capital cost . . . it is proposed that the percentage in respect of engines and carriages should be raised from 2¼% per annum to 2½% per annum, commencing with the current half-year, that is the half-year ending 31st January 1912.

At this time there were 927 engines valued at £1,858,519.

From 1909 more money became available for the construction of new locomotives. Had the extra money been made available a year earlier, it is conceivable that the Cardean class might have been increased or an alternative design produced. But by the autumn of 1909, it was evident that the firetube superheater provided an excellent method of increasing the capacity of a locomotive at relatively little additional cost. Rather than pursue a 'large engine' policy, with the possibility of having to spend money on development, McIntosh decided that the best way to obtain the maximum benefit from the extra capital was to build superheated versions of his well-tried, medium-sized engines.

THE 139, 117 AND 40 CLASSES

In 1909 an order was placed for four additional engines of the 140 class. The drawing office updated the design, new drawings being prepared for such details as the coupled wheels and the axlebox guides. The spark arrestor and the built-up crank-axle were by now standard fittings and the draughting arrangements were altered in the light of the 1909 trials. But the most noticeable change lay in the design of the tender. The tank was the same size as that of the big 4–6–0s but the frames and bogies were completely redesigned, the well being abandoned and the bogies sprung by individual laminated springs above the axleboxes. The last engine of the batch was given a Schmidt superheater and piston valves and, when completed in July 1910, was the first engine in Scotland to have high-degree superheat.

THE COMING OF SUPERHEATING

Order No	Engine Nos	Dates delivered	Cost per engine
Y92	137-8, 136,	6/10—7/10	£2540
	139	7/10	£2956

The engines were dual-fitted.

The 24-element superheater fitted to No 139 was of the Schmidt type, the elements being bolted to the header. Each passed backwards within a 5in diameter flue to a point roughly half-way between the dome and the firebox tubeplate and, before returning to the header, formed a loop extending forward into the smokebox. These 'long return loop' elements had a very high surface area but the official figure for the superheating surface, 330sq ft, was the area of the *external* surface of the elements; soon after, engineers calculated superheating surface from the *internal* surface of the elements. To prevent the elements from burning when the regulator was closed, a large damper was pivoted from the header. The angle of the damper could be altered by the application of steam to a cylinder on the right-hand side of the smokebox, this in turn being controlled by a rod running backwards above the handrail and terminating in a brass wheel just in front of the cab 'lean out'. The piston was connected to a short weighshaft within the smokebox, with lifting arm on one side and balance weight on the other. When the lower end of the damper was in its most backward position, gaseous flow through the superheater flues was shut off. The presence of the superheater header, damper and control gear precluded use of the complete spark arrestor though louvres

TABLE 7
139 CLASS VALVE SETTING: FORWARD GEAR

Notch	Travel, in	Total lead, in* FP	BP	Port opening, in FP	BP	Cut off, % FP	BP	Release, % FP	BP
9	$4\frac{1}{8}$	$\frac{1}{8}$	$\frac{1}{16}$	$1\frac{1}{32}$	$1\frac{5}{32}$	78	76	93	91
8	$3\frac{7}{8}$	$\frac{3}{16}$	$\frac{1}{8}$	$\frac{7}{8}$F	1B	75	71	91	89
7	$3\frac{5}{16}$B	$\frac{1}{4}$	$\frac{5}{32}$	$\frac{3}{4}$B	$1\frac{3}{8}$B	68	66	88	86
6	$3\frac{1}{4}$F	$\frac{5}{16}$	$\frac{3}{16}$	$\frac{5}{8}$	$\frac{5}{8}$	62	60	86	83
5	3	$\frac{3}{8}$	$\frac{7}{32}$	$\frac{1}{2}$	$\frac{1}{2}$B	53	52	82	79
4	$2\frac{3}{4}$F	$\frac{13}{32}$	$\frac{1}{4}$	$\frac{3}{8}$F	$\frac{3}{8}$	42	42	76	74
3	$2\frac{9}{16}$F	$\frac{7}{16}$	$\frac{1}{4}$	$\frac{5}{16}$F	$\frac{1}{4}$	32	32	71	68
2	$2\frac{15}{32}$	$\frac{15}{32}$	$\frac{9}{32}$	$\frac{9}{32}$	$\frac{3}{16}$	22	20	62	60
1	$2\frac{9}{32}$	$\frac{1}{2}$	$\frac{5}{16}$	$\frac{1}{4}$	$\frac{5}{32}$	15	12	53	52

FP=front port   BP=back port
F=full   B=bare
* exactly half the total lead was derived from the Trick port

THE COMING OF SUPERHEATING

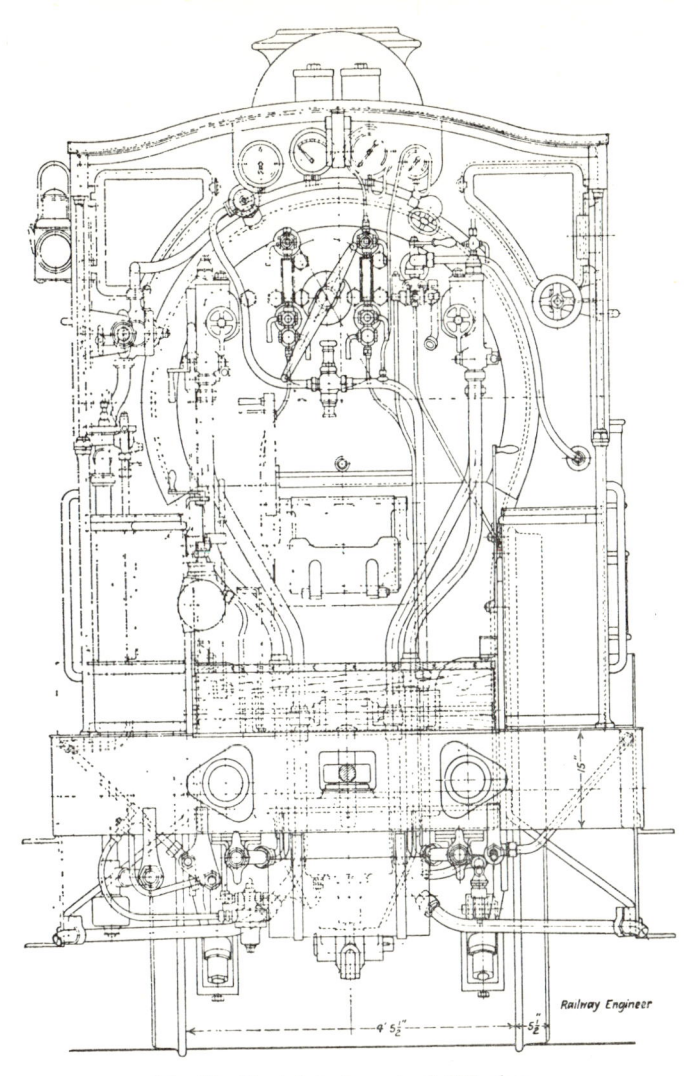

Fig 19　Footplate layout of 177 class

were fitted to the inside of the smokebox door. As it was, the smokebox was lengthened by $8\frac{5}{8}$in.

Working pressure was fixed at 165lb/sq in to reduce boiler maintenance costs and the diameter of the cylinders enlarged to 20in. The steamchests were above the cylinders, steam distribution being by inside-admission 8in diameter piston valves with single wide Schmidt rings. The valves were driven indirectly via rocking levers, the eccentric-rods being only 4ft 6in long. The valve events are shown in Table 7. A snifting valve attached to the waist of the smokebox on either side communicated with the adjacent steam-chest. The built-up crank-axle was similar to that of the 918 and 908 classes though the wheel-seats were larger. The frames were 1$\frac{1}{16}$in thick and laminated springs were used for both coupled axles. Front-end lubrication was by a Friedmann mechanical lubricator, mounted on the right-hand running-plate and driven from the slide-block of the crosshead. Footplate fittings included a pyrometer, attached to the right-hand side-sheet, two No 9 Gresham & Craven combination injectors and the standard steam-assisted reverser, with nine notches on either side of mid-gear.

Nos 138-9 were sent to Perth and ran trials on the Carlisle road. Each engine made three return trips, travelling south on the 9pm sleeper and returning on the 2.20am Tourist:—

Engine No		138	139
Average trailing load, tons: up		265	220
	down	220	235
Total coal, lb		42,292	33,520
Coal per train-mile, lb		47.5	37.2
Coal per gross ton-mile, lb		0.138	0.106
Total water, gallons		27,495	22,138

There was a saving of 23$\frac{1}{2}$ per cent in coal and 25$\frac{1}{2}$ per cent in water for the superheated engine. No 139 was tested also on the 10am Grampian Corridor Express from Buchanan Street to Perth, indicator diagrams being taken at 14 points. For the easier sections, the reverser was put in the second notch, this giving 21 per cent cut off, and it was on this setting that the maximum power output was attained—940ihp at 50mph. But while the diagrams themselves showed excellent characteristics, no attempt was made to extend the engine to its limits. With only 205 tons behind the tender, the climb from Bridge of Allan to Kinbuck was made with the lever in the third notch, equivalent to a cut-off of 32 per cent, and speed fell away to a minimum of 21$\frac{1}{2}$mph.

These tests showed that the very economical running of the superheated engine was largely due to the very high steam temperature (670°F) reached. Although the maximum tractive capacity of the engine had not been established, a pointer had been

Page 141 (*upper*) 60 class 4–6–0 No 62 on down express near Rockcliffe; (*lower*) 4–6–2T No 944 on Gourock train near Ibrox

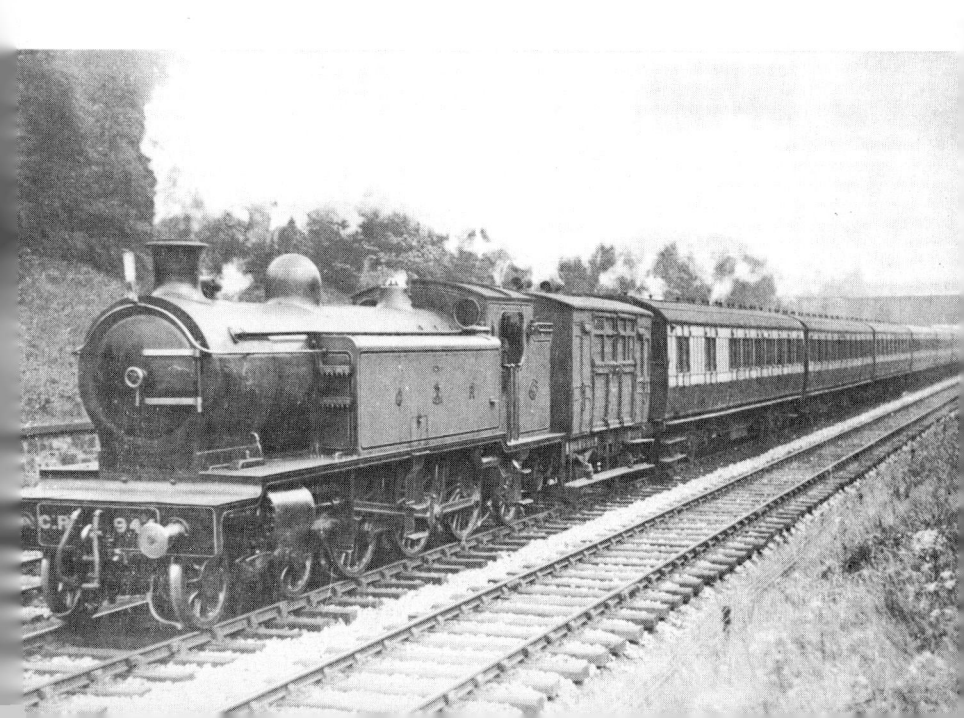

Page 142 (upper) 956 class 4-6-0 No 959 near Stanley Junction; *(lower)* 191 class 4-6-0 No 192 near Bridge of Allan

THE COMING OF SUPERHEATING

obtained to the economy which would result from a general application of the principle. Over the next few years, orders were placed for four batches of 4-4-0s generally similar to No 139:—

Order No	Engine Nos	Dates delivered	Cost per engine
Y97	132-5	4/11—5/11	£2884
Y101	117-9, 120-2	7/12—8/12	£2830 & £2845
Y105	43-8	5/13—6/13	£2762
Y109	41-2, 39-40, 123	4/14—5/14	£2802

Nos 132-5 were fitted with Wakefield No 1 mechanical lubricators but were otherwise identical to No 139. In batch Y101, the diameter of the cylinders was increased to 20½in (*not* 20¼in as has often been stated) and the bogie axles and axleboxes altered slightly. Nos 117-121 were fitted with the steam-operated superheater damper but No 122 got a Robinson draught retarder, confusingly referred to at St Rollox as a 'steam damper'. The cylinders of Nos 120-2 were fitted with pressure-relief valves. Nos 120-1 were equipped with Stone's mechanical lubricator.

Batches Y105 and Y109 formed a sub-class which combined 20¼in cylinders with what was known at St Rollox as the 'Consolidated' superheater. This was of the Robinson type, the elements being expanded into the header, but protection against burning was afforded by a hand-operated damper, the control-rod running back alongside the boiler to a lever and quadrant within the cab. The rear of the elements lay slightly further forward of the firebox than before, superheating surface being reduced to 295sq ft. Nos 39-42, 48 and 123 were fitted with Stone's mechanical lubricator; the remainder probably received one of Wakefield's products. The engines were dual-fitted and coupled to the 4600-gallon pattern of bogie tender introduced with batch Y92. They were allocated as follows:—

    Aberdeen Nos 40-4, 117-8
    Carlisle Nos 47-8, 121-3, 134-5
    Perth Nos 39, 45-6, 119-20, 132-3

Two turns which fell to the Carlisle and Aberdeen engines were the 6.40am Aberdeen to Carlisle, returning with the 2.54am down Postal, and the 1.12pm from Carlisle to Aberdeen, returning with the 3.40pm up Postal, the low coal and water consumption of the superheated engines making them eminently suitable for such lengthy runs. These were worked by Carlisle and Aberdeen men

THE COMING OF SUPERHEATING

on alternate days. On the down Tourist, it was not unusual for the Perth engines to run the 150 miles from Carlisle to Perth without taking water. Water consumption cannot have been greater than 30 gallons a mile. The schedule, however,—143min for the 117¾ miles from Carlisle to Stirling—was much easier than that in force during the palmy days of 1896. When loads approached the 400-ton mark, a pilot was provided as far as Beattock Summit.

SUPERHEATED GOODS ENGINES

Extension of superheating to goods classes began in 1912 when four 0–6–0s were constructed at St Rollox:—

Order No	Engine Nos	Dates delivered	Cost per engine
Y98	30-3	7/12—11/12	£2312

These engines were based on the 652 class but their boilers were pitched 6in higher and contained a 21-element Schmidt superheater with long return loops. Again the elements were protected by a steam-operated damper. The chimney and blastpipe were arranged in accordance with the results of the 1909 trials and louvres fitted to the inside of the smokebox door. The smokebox was lengthened by 1ft 2¼in and the front overhang by 2ft 3in. This increase was necessary to house the casing for the long piston tail-rods. The valve-chests were placed on top of the cylinders which were brought 3in closer together and enlarged to 19½in in diameter. Steam distribution was by 8in diameter inside-admission piston valves with the single wide Schmidt ring and Trick ports. The valve spindles were driven indirectly via rocking levers, this

TABLE 8
30 CLASS VALVE SETTING: FORWARD GEAR

Notch	Travel, in	Total lead, in* FP	BP	Port opening, in FP	BP	Cut off, % FP	BP	Release, % FP	BP
9	4 5/16	13/32	1/16	1 1/8 B	1 3/16	80	75	94	89½
8	4	15/32	1/8	31/32	1 1/32	76	71	92	87
7	3 11/16	½	3/16	13/16	7/8	71	66	90	85
6	3 7/16	9/16	¼	11/16	¾	64	60	87	82
5	3 5/32	19/32	9/32	9/16 F	19/32	55	53	83	78
4	2 31/32	5/8	11/32	15/32	½ B	45	45	78	74
3	2 25/32	5/8	3/8	13/32	3/8	35	36	72	68
2	2 5/8	5/8	13/32	11/32	9/32	26	27	64½	62
1	2 9/16	5/8	13/32	5/16	¼	18	19	56½	55

Lap of valve 1in
* exactly half the total lead was derived from the Trick port

THE COMING OF SUPERHEATING

arrangement requiring eccentric-rods only 4ft 2in long. The valve events are shown in Table 8. Because of the increase in weight, particularly at the front of the engine, provision for longer journals was made by dishing the wheels. The dimensions of the journals were:—

	dia, in	length, in
Leading	$8\frac{1}{2}$	$9\frac{1}{4}$
Driving	$8\frac{1}{2}$	9
Trailing	8	$7\frac{3}{4}$

The crank-axle was of the built-up pattern with parallel-sided webs. Reverse was by unassisted lever and front-end lubrication by a mechanical lubricator driven off the slide-block of the crosshead.

Although officially designated goods engines, the locomotives were Westinghouse-fitted and painted blue for service on the Glasgow Central—Wemyss Bay route, two being allocated to Polmadie and two to Greenock. With a tractive effort of 22,409lb and an adhesion weight of 51 tons $2\frac{1}{2}$cwt, they proved ideal engines for the route.

Before building further engines of the class, McIntosh decided to reduce the weight on the leading coupled wheels by lengthening the front overhang and supporting it on a pony truck. Five engines thus modified were built at St Rollox towards the end of 1912:—

Order No	Engine Nos	Dates delivered	Cost per engine
Y103	34-8	11/12—12/12	£2329

The frames were lengthened by 2ft 6in and braced by a heavy cross-stay from which the pony truck was pivoted. Lateral movement was controlled by helical springs compressed between the pony truck and the main frames of the engine. In most other respects the new engines resembled their immediate predecessors but the superheater damper was omitted and the steam brake fitted instead of the Westinghouse. A vacuum ejector and through pipes were provided for the operation of vacuum-braked goods trains. The engines were painted black, allocated to Carlisle and employed on the following duties:—

(1) 1.40pm Carlisle to Perth   (3) 1.30am Carlisle to Dundee
    1.15pm Perth to Carlisle   9.35pm Dundee to Carlisle
(2) 9.00pm Carlisle to Perth   (4) 4.25am Carlisle to Dundee
    4.50pm Perth to Carlisle   Return as required

THE COMING OF SUPERHEATING

These turns were worked on alternate days by Perth engines. The spare 2-6-0 and crew were allowed a week of shorter-distance work, generally to Carstairs, between spells on the long runs.

Although the 2-6-0s proved to be most capable engines, it was realised that something larger was required for the more important fitted goods trains. A superheated version of the 908 class was designed, the first batch being turned out as follows:—

Order No	Engine Nos	Dates delivered	Cost per engine
Y107	179-83	12/13—1/14	£2900

The boiler contained a 24-element 'Consolidated' superheater with long return loops, the hand-controlled damper being operated from a quadrant and lever within the cab. The valve-chests were placed on top of the cylinders which were brought 3in closer together and enlarged to 20in dia. Steam distribution was by 9in dia standard piston valves, driven via rocking levers, the eccentric-rods being 4ft 2in long. Front-end lubrication was by a Wakefield 10-feed mechanical lubricator driven from the cross-head. The frames were $1\frac{1}{8}$in thick and the built-up crank-axle re-designed to suit the altered disposition of the cylinders, the journals being lengthened to 9in. The intermediate and trailing axles had concave journals 12in long and were allowed a side-play of $\frac{1}{8}$in. The blastpipe and chimney were arranged in accordance with the results of the 1909 trials (p.135), with louvres fitted to the inside of the smokebox door. From a mechanical point of view, the new engines were otherwise very similar to their saturated predecessors but, with much longer smokeboxes and side-window cabs, were distinctive in appearance. (The side-window cab had been fitted experimentally to No 917 in 1907.) Although painted blue, the 179s were officially designated goods engines and lacked the special distinguishing mark of express engines, the gilt scrolls on either side of the tender coat of arms. Nevertheless, No 179 frequently worked the 10am Grampian Corridor Express between Buchanan Street and Perth during the summer of 1914.

In 1914 the class was further expanded:—

Order No	Engine Nos	Dates delivered	Cost per engine
Y112	184-9	12/14—3/15	£2749

These engines differed from those of the first batch in having $19\frac{1}{2}$in diameter cylinders. All had vacuum ejectors and brake

THE COMING OF SUPERHEATING

valves coupled to the Westinghouse brake system in the usual way.

When complete, the class was distributed as follows:—

        Carlisle   Nos 179, 188-9
        St Rollox   Nos 180-3
        Perth   Nos 184-7

Two of the Carlisle engines were put on 'book-off' goods working to Perth while the remaining engine ran to Dundee in a link shared with a 2–6–0. The St Rollox quartet worked the 5.30am and 7.45pm to Carlisle and the 1.30pm and 11.55pm to Aberdeen, but during the Glasgow Fair holidays they helped out with passenger work. Of the Perth engines, Nos 184-5 worked for a time on express passenger duties, the engines appearing in alternate weeks on the 10am and 5pm from Buchanan Street to Perth. Later all four were on 'book-off' goods rosters to Carlisle. What this involved is shown in Table 9. Although handicapped by the small, shallow grate and restricted ashpan space of their saturated forebears, the 179s were reasonably economical engines and were well liked by the men. Coal consumption on the Glasgow-Aberdeen

TABLE 9
179 CLASS WORKINGS

Date		14/11/17			15/7/18	
Engine No		188			189	
	Arrive	Depart	Wagons	Arrive	Depart	Wagons
Dundee	—	9.58pm	40			
Perth General	10.45	10.55	40			
Perth South	11.00	11.25	23			
Cornton	12.35	12.40	23Sigs			
Stirling	12.48	1.00am	31	—	7-10pm	44
Polmaise				7.20	7.55	48
Greenhill	1.33am	1.45	31Water	8.30	8.35	48
Mossend				9.10	9.25	48
Shieldmuir				9.55	10.30	48
Law Jtc	2.40	2.48	39			
Braidwood				11.10	11.35	48
Carstairs	3.25	3.35	39	12.00	12.15am	48
Crawford				12.43am	1.05am	48Sigs
Beattock Summit	4.35	4.40	39	1.40	1.55	48
Beattock	5.03	5.13	39	2.10	2.20	48
Wamphray				2.50	3.00	48
Kirtlebridge	5.50	5.56	39			
Kirkpatrick	6.02	6.32	39Water			
Gretna	6.40	7.00	39			
Floriston				4.08	4.20	48
Kingmoor	7.15	—		4.37	—	

THE COMING OF SUPERHEATING

goods trains is said to have been about four tons, which would mean something in the region of 55-60lb per mile.

RE-EQUIPMENT

With a potential fuel economy of some 20 per cent in mind, McIntosh rebuilt the 6ft 6in 4–6–0s with superheaters and piston valves. The seven engines were dealt with as follows:—

Engine Nos	Date	Engine Nos	Date	Engine Nos	Date
907	2/11	49	3/11	50, 906	4/11
903-4	5/11	905	7/11		

Since all retained their original frames, they were not officially designated 'rebuilds'. The 24-element superheaters were of the Schmidt pattern with long return loops and a steam-operated damper. The header was placed within the smokebox proper, rather than against the tubeplate. Although length between tubeplates remained unchanged, 15ft 8in for Nos 49-50 and 16ft 8in for the remainder, the same length of element was used for all seven engines, the return loops of the 49 class projecting some 15in beyond the mouths of the flues! Steam temperature was recorded by a pyrometer attached to the left-hand side-sheet of the cab. Working pressure was reduced to 175lb/sq in and new 20¾in dia cylinders fitted. Steam distribution was by 8in dia piston valves with a single wide Schmidt ring and Trick ports, the change from outside to inside admission necessitating a reversal in the positions of the eccentric pulleys. Front-end lubrication was by a Wakefield eight-feed mechanical lubricator driven off the crosshead.

On their return to traffic, the 4–6–0s resumed their former duties though No 49, which is said to have been very susceptible to overheating, was soon put on a Glasgow-Carlisle fitted goods. About this time, James Currie was promoted to inspector and his place on *Cardean* taken by David Gibson. Thus began an association which became something of a legend though it is doubtful if Gibson ever managed to surpass the work of his predecessor. Superheating made very little difference to the day-to-day performance of the engines though the coal consumption was greatly reduced; whereas a return trip on the Corridor previously would have almost stripped the tender, it was now common to have about one-third of the coal left at the end of the day. *Cardean* continued on the Corridor until 1916. About 1910, she had been fitted with

THE COMING OF SUPERHEATING

a whistle of very large diameter and her unusually deep-toned hoot could be heard for miles.

In March 1913 the drawing office finished work on the superheating of the 766 and 900 classes but it was not until the following year that rebuilding began. Six engines were dealt with:—

Engine Nos	Date	Engine Nos	Date
771 & 901	3/14	772 & 898	7/14
766	10/14	769	12/14

No 898 kept her original frames and was not officially designated a rebuild; the other five got new frames and were classed as rebuilds. The new frames were modified to match the new, longer smokeboxes. The boilers of the rebuilt 766 class engines were pitched 3in higher than before to bring them into line with those of the 900 class. The boilers were fitted with 18-element 'Consolidated' superheaters with long return loops and a hand-controlled damper. Working pressure was reduced to 170lb/sq in and new 19½in dia cylinders fitted. Although of the same diameter as those of the 30 class, the cylinders were spaced 3in further apart so that the existing crank-axles could be retained. Steam distribution was by 8in dia standard piston valves, positioned above the cylinders and driven via rocking levers, the indirect drive necessitating a change to 4ft 6in eccentric-rods. Front-end lubrication was by an eight-feed mechanical lubricator driven from the crosshead.

After rebuilding, No 771 was sent to Perth, No 898 to Carlisle and the remainder to Polmadie for the crack Gourock and Ardrossan boat trains. About a year later No 772 won renown while deputising for *Cardean* on the down Corridor, loaded to 395 tons gross. C. J. Allen was a passenger that day and he recorded the result in *The Railway Magazine* of October 1915. The highlight was the sustained minimum of 44mph on the long stretches of 1 in 200 north of Gretna, an effort which must have involved an output of 955edhp. The 39.7 miles to Beattock were run in 47min 15sec, start to stop.

It was not until May 1915 that the first engine of the 140 class, No 924, was superheated. Another scheme of this period, but one which never came to fruition, was a scheme to rebuild the 908 class 4–6–0s with 24-element superheaters. Patent metallic packing designed at St Rollox in 1913 was used for the cylinders and valve chests of superheated engines.

THE COMING OF SUPERHEATING

From 1911 onwards, steam heating apparatus was gradually applied to the Westinghouse-fitted 0–6–0s. During the miners' strike of 1912, one Dunalastair—No 724—and two 812s—Nos 285 and 292—were temporarily converted to burn oil-fuel on the Holden system. The oil was carried in two cylindrical tanks, each of 260 gallons capacity, placed alongside each other in what was normally the coal-space. In June 1913, a built-up crank-axle was introduced for use on the 18in x 26in passenger, goods and mineral engines. In the same year, it was planned to fit 140 class engine No 136 with a Weir feed-water heater and pump. The apparatus was carried on the left-hand running-plate, the feed entering the bottom of the boiler. Soon, however, top-feed was substituted, the clack-box being mounted behind the dome. The apparatus was removed in 1915 but the experiment was resumed in 1920 when a new heat exchanger and feed-pump, both of modified design, were fitted in conjunction with top-feed.

TANK ENGINES: 1910-14

The only new design brought out during this period was an outside cylinder 0–6–0T, suitable for lines too sharply curved for the 782 class engines. Initially, only two engines were built:—

Order No	Engine Nos	Dates delivered	Cost per engine
Y100	498-9	1/12	£1219

They were a big improvement on the 0–4–0STs but did not have quite the same scope:—

Class	Tractive effort	Wheelbase	Minimum curvature
262	10,601lb	7ft	1½ chains
498	18,014lb	10ft	3 chains
782	19,890lb	16ft 3in	4½ chains

Although outside the frames, the 17in by 22in cylinders were reminiscent of Drummond practice, their front and rear covers being contoured to accommodate typical Drummond cone-shaped pistons. The big-ends were in one piece, bushed for the crank-pins and extended upwards to house an oil reservoir. The valve-chests were inside the frames, the slide-valves being driven by Stephenson link motion. Since the expansion-links were behind the leading axle and the cylinders well in front, each valve-spindle was con-

THE COMING OF SUPERHEATING

nected to its die-block by means of an intermediate link shaped like an inverted U, the axle passing between the two arms, the lower ends of which were united by a stiffening-plate. The valve events are shown in Table 10. The boiler barrel was of the same

TABLE 10
498 CLASS VALVE SETTING: FORWARD GEAR

Notch	Travel, in	Lead, in FP	Lead, in BP	Port opening, in FP	Port opening, in BP	Cut off, % FP	Cut off, % BP	Release, % FP	Release, % BP
6	$3\tfrac{29}{32}$	$\tfrac{3}{16}$	$\tfrac{3}{32}$	$\tfrac{29}{32}$	1	75	71	93	90
5	$3\tfrac{9}{16}$	$\tfrac{7}{32}$	$\tfrac{3}{32}$	$\tfrac{3}{4}$	$\tfrac{13}{16}$	68	64	90	86
4	$3\tfrac{1}{4}$	$\tfrac{1}{4}$	$\tfrac{5}{32}$	$\tfrac{19}{32}$	$\tfrac{21}{32}$	58	56	87	82
3	$2\tfrac{15}{16}$	$\tfrac{9}{32}$	$\tfrac{3}{16}$	$\tfrac{15}{32}$	$\tfrac{15}{32}$	48	46	82	78
2	$2\tfrac{1}{4}$	$\tfrac{9}{32}$	$\tfrac{3}{16}$	$\tfrac{13}{32}$	$\tfrac{11}{32}$	36	35	75	71
1	$2\tfrac{19}{32}$	$\tfrac{5}{16}$	$\tfrac{7}{32}$	$\tfrac{11}{32}$	$\tfrac{1}{4}$	24	23	65	63

Lap of valve 1in

diameter as that of the 782 class but was $9\tfrac{1}{2}$in shorter and pitched 2in lower. The outer firebox was 8in shorter than that of the inside-cylinder engines and was approximately 1ft less in depth throughout. The hood was of the 1909 pattern, its conical mouth terminating 7in above the blastpipe orifice, which was on a level with the second top row of tubes. Suspension was by laminated springs below the axleboxes. The brake-rigging was compensated and was operated by steam in the usual way. Water capacity was 1000 gallons and coal 2 tons.

No 498 was allocated to St Rollox and was employed mainly in shunting Braby's Eclipse works close by, while No 499 went to Grangemouth for dockyard duties.

Further 0–6–0Ts of the 782 class were added to stock as follows:—

Order No	Engine Nos	Dates delivered	Cost per engine
Y91	432-3, 475	1/10–2/10	£1250
Y93	435-6, 476-80	7/10–8/10	£1250
Y95	416, 418, 481	3/10	£1382
	474, 483-4, 500, 608-10	10/10–11/10	
Y99	277, 390, 393, 397, 405, 430 471	4/11–10/11	£1379
Y104	651, 781, 130-1	10/12–12/12	£1398
	171-4	10/12–12/12	£1450
Y108	175-8, 508-9	9/13–10/13	£1392

THE COMING OF SUPERHEATING

Nos 171-4 were Westinghouse-fitted though painted black. No 436 was fitted with a vacuum ejector and sent to Aberdeen for shunting fish vans.

By the end of 1913, the *approximate* numbers allocated to each shed were as follows:—

Polmadie	27	Motherwell	23	Hamilton	20
Grangemouth	10	Dawsholm	9	Dalry Rd	5
Carstairs	4	Carlisle	4	Perth	4
St Rollox	3	Stirling	3	Aberdeen	1

The principal rôle of these engines was working mineral and goods trains in the industrial belt; their rôle in shunting the yards at the main provincial centres was by comparison a minor one.

Standard 0-4-4Ts were constructed as follows:—

Order No	Engine Nos	Dates delivered	Cost per engine
Y94	155, 160, 459, 461	4/10−1/11	£1582
Y96	152-4, 156, 380, 462	3/11−7/11	£1580
Y102	157, 383, 457-8	4/12−7/12	£1576
Y110	222-7	8/14−10/14	£1543

By the time that batch Y110 was complete, the *approximate* numbers of standard 0-4-4Ts allocated to each area were as follows:—

Polmadie	12	Ardrossan/			
Motherwell	6	Kilwinning	2		
Perth	7	Crieff	2		
Dundee	4	Blairgowrie	1		
Forfar	7	Brechin	3	Montrose	2
Beattock	5	Leadhills	1	Lockerbie	3
Carstairs	3				
Dalry Rd	3				
Stirling	2				
Aberdeen	1	Arbroath	1		
Dawsholm	1				

Of the Forfar group, one engine was reserved for the Alyth branch. The class had an almost complete monopoly of the branches serving Angus and Eastern Perthshire. In the south its stronghold was at Beattock, which provided engines for the Leadhills, Moffat and Dumfries branches as well as pilot engines for banking trains to Beattock Summit.

THE COMING OF SUPERHEATING

ASSESSMENT

On 11 October 1911, in a short ceremony aboard the Royal Train, McIntosh was invested as a member of the Royal Victorian Order. It was a fitting recognition of the part he had played in locomotive development for, until about 1904, his 'big engine' policy had kept his company in the van of railway progress, innovation and enlargement being matched by superb performances on the road. But from about 1908, St Rollox's reputation began to wane. It was not just a question of the drawing office resting on its laurels, or of financial restrictions. With some notable exceptions, there was no longer the same zest in locomotive running. Uphill work was still good but downhill running tended to be restrained, even when there was time to be regained. Piloting was on the increase and in 1908 McIntosh issued the following instructions:—

> It has been noticed lately that there has been an increase in piloting express trains especially West Coast through trains. In some instances the reason for this step is not at all clear and Locomotive Foremen and all concerned are requested to examine carefully all requests for pilot assistance. It is obvious that at times when West Coast trains are late in arrival at Carlisle, it has been considered necessary to attach a second engine. This of course, unless the load requires extra assistance, should not be allowed to continue.

While the remedy for the apparent decline in the standards of performance may have rested largely with the men, it is likely that performance suffered also from a number of technical imperfections. The relatively small grate areas meant that heavy demands on the boiler could be met only by a disproportionately high rate of combustion. The spark arrester may have had some adverse effect on steaming and the altered chimney and blastpipe arrangements which accompanied it were very far from the theoretical ideal. Moreover, superheating was not the panacea first supposed, for it introduced its own special problems, carbonisation of oil in the steam passages perhaps being the most important.

Pragmatic rather than inventive, McIntosh explained his basic philosophy in an article published by *Cassier's Magazine* in 1910. 'Efficiency and reliability are more to a locomotive superintendent than economy in fuel, and those who provide public facilities are wise not to take risks for the sake of small gains'. A locomotive superintendent's main function was not to perfect the locomotive; it was to manage a large body of men engaged in a wide variety of activities, to make sure that the capital invested was used efficiently, to knit the whole into one unit and to keep costs to the minimum. He had much routine administrative work to do and a lot of time had to be spent in correspondence with the general manager and the various board committees. A high professional reputation provided no immunity against the receipt of irksome letters from financial directors who, despite their salary of £5250 per annum, sometimes had little understanding of railway operation. On 29 September 1908, the finance committee ordered McIntosh to explain why, after decreases of 3661 and 3454 tons in the quantity of coal used during the previous two fortnights, the reduction during the first fortnight of September was only 2435 tons. In such a climate it was wiser to keep to the well-tried ways of Drummond than to launch into experiment.

McIntosh tended to view locomotive design from the standpoint of the engineman rather than from that of the engineer. In *Cassier's Magazine*, he wrote:—

> The outside-cylinder six-coupled engine with its long connecting-rod and direct thrust on the coupling-rods, its direct motion and large driving bearings, has apparently some decided advantages over the inside-cylinder type, with its greater frictional resistance in the driving axle-box, its rocking-shaft motion, and comparatively short connecting-rod. Each design has its supporters, but this much may be said for the inside-cylinder, that the greater steadiness in running, due to the closeness of the centres of the cylinders, probably balances all its other defects.

Curiously enough, an outside-cylinder 4–6–0 was designed at St Rollox in 1911. It was to have had 6ft 6in coupled wheels and 21in by 26in cylinders. That the construction of this engine was seriously considered, and not just a draughtsman's pipe-dream, is shown by the fact that detailed drawings of the cylinders, valve-chests and various other components were completed. Although put aside, the project eventually saw the light of day at a later date. A proposal for an outside-cylinder 4–6–0 with 5ft 9in wheels was also put forward about this time, though no detailed drawings for this engine seem to have been made.

McIntosh's last big project, and one which demonstrated his breadth of vision, came in the autumn of 1913 and was for a four-cylinder Pacific, the proposed dimensions of which are shown in Appedix 7. While criticism may be levelled at the very long boiler-barrel and the shallow grate, the design offered more scope for development than did that of the inside-cylinder 4–6–0s.

CHAPTER 10

WILLIAM PICKERSGILL

METAMORPHOSIS

Towards the end of 1913, McIntosh intimated his wish to retire and on 9 December the board resolved to appoint William Pickersgill in his place, the relevant minute reading:—

> Locomotive Superintendent, Resignation of Mr J. F. McIntosh. Agree to appoint Mr William Pickersgill as Locomotive Superintendent at a salary of £1250 per annum from a date in May 1914 to be fixed by agreement.

The choice was surprising; for 20 years, Pickersgill had been the locomotive superintendent of the Great North of Scotland Railway, whose largest engines were no bigger than the Drummond 4-4-0s of 1884. He took up office on 4 May 1914 though McIntosh remained on hand until 28 May to ensure a smooth changeover. Both men attended the locomotive and stores committee on 18 May, McIntosh for the last and Pickersgill for the first time.

William Urie also retired on 28 May, his duties passing to Irvine Kempt Junior who was given the titles 'assistant locomotive superintendent (works)' and 'chief inspector'. John G. Barr became 'assistant locomotive superintendent (running)' with the title 'inspector'. Both men were paid £450 per annum.

Pickersgill had been in office only four months when World War I broke out and it was not long before the locomotive department was facing its biggest crisis since 1882. His major concern was the loss of skilled men to the armed forces.

The rise in traffic as the war effort gathered momentum further aggravated the situation. Traffic to and from the steelworks and heavy engineering establishments around Glasgow soared and passenger trains could no longer be given priority. The express services were drastically decelerated, thus enabling very heavy loads to be run with one engine. By the end of the war, the superheated engines, 4–4–0s and 4–6–0s alike, were taking 440 tons tare south of Carstairs without a pilot. Such work led to an increase in maintenance but there just weren't enough skilled men to carry out the repairs. On 22 February 1916 the following minute was recorded:—

> Due to the scarcity of Fitters and Erectors, it is impossible to properly overtake the repair of locomotive engines in the sheds at St Rollox. It is therefore suggested that endeavour should be made to get a number of engines repaired by outside contractors. The Locomotive Superintendent has been in communication with locomotive builders and reports that the following three firms are prepared to undertake the work on the terms stated:— Robert Stephenson & Co., . . . R. & W. Hawthorn Leslie & Co., . . . and The Yorkshire Engine Co., . . .
> Remit to General Manager with powers to arrange for repairs as proposed.

The shortage of labour also affected locomotive construction. And for the first time since 1899, tenders were invited from the locomotive-building industry. Indeed, so serious was the position that in October 1917, an order was placed with the North British Locomotive Company for the machining of 16 engine frame plates. This meant a considerable increase in expenditure at a time when the cost of many other essential commodities was rising sharply; from about 10s per ton in May 1914, the price of locomotive coal had risen to 19s per ton by March 1915.

Another source of concern was the loss of locomotives to other railways. In 1915 the hard-pressed Highland Railway obtained temporary possession of four six-coupled tender engines, Nos 56, 314, 555 and 560, together with six Drummond 4–4–0s, Nos 70, 73, 75-6, 89 and 91. More serious was the acquisition of 25 'Jumbos' by the War Department for service overseas. To help balance the loss four Fletcher 398 class 0–6–0s of the North Eastern Railway (Nos 282-3, 308 and 1378) and seven Great Central Railway 0–6–0s (Nos 796, 801, 809, 826, 830 and 835-6) were transferred to the Caledonian.

THE RIVER CLASS

In 1915 the Caledonian received an unexpected windfall of six new mixed-traffic 4–6–0s of advanced design. Not long after his appointment as locomotive superintendent of the Highland Railway, Frederick Godfrey Smith had decided to build six 4–6–0s of greater power than any hitherto run on the Highland line and an order was placed with R. & W. Hawthorn Leslie & Co Ltd. But although the detailed specification was issued from Lochgorm works and bore Smith's name, some, if not all, of the drawings were prepared as a war-time expedient in the North British Railway drawing office at Cowlairs by Archibald Campbell, later chief draughtsman of the NBR, various NBR features, such as drop grates and hot water injectors, being incorporated in the design.

On 24 September 1915, a special meeting of the Highland Railway board was convened, the purpose and deliberations of the meeting being recorded in the minutes.

> The Deputy Chairman of the Company (Mr Whitelaw) explained that the meeting had been called to consider the situation which had arisen owing to the fact that the six new Engines presently being built by Messrs Hawthorn Leslie & Co. and one of which has been delivered at Perth were now found to be considerably heavier than the Directors anticipated. Mr Whitelaw read correspondence with the contractors, explained the arrangements between the Directors and Mr Smith, the Company's Locomotive Superintendent, as to the design of the Engines and made a statement as to Engine weights on Bridges, etc. Mr Smith was then given the fullest opportunity of stating his case.
>
> After full consideration, the Board resolved that an opportunity should be given to Mr Smith to resign his appointment forthwith, and that his resignation, if tendered, will be accepted and three months' salary paid to him; further, if his resignation is not received by Monday next, his engagement will be forthwith terminated and three months' salary paid to him in lieu of notice.

On 5 October, a proposal to purchase the six engines was considered by the Caledonian Railway's traffic committee:—

> Submitted memorandum from the General Manager in regard to a proposal to purchase six new engines from the Highland Railway Company.
> Authorise General Manager to offer up to £5,500 per engine for the six engines in question. Capital.

On 7 October, the Highland Railway board 'Resolved to sell the six new engines to the Caledonian Railway Company at the price of £5400 each, payment being made on delivery of each engine'. Hawthorn's price per engine was £4920. The locomotive famine made it possible for the Highland to demand a quick sale and a sizeable profit. On 4 November the HR board

WILLIAM PICKERSGILL

Resolved to inform the Caledonian Railway Company that we intend to pay for the first two Engines next week; that delivery of the remaining Engines will be made at Carlisle or Perth on condition that payment is made on delivery; and that unless this arrangement is agreed to within two days after next meeting of the Caledonian Board, the Highland Company will hold themselves free to accept another offer.

The deal was concluded on 16 November when a final price of £32,400 was agreed.

In appearance as well as in general design, the new 4–6–0s were unlike anything previously seen in Scotland. Their 21in by 28in outside cylinders were inclined at 1 in 45 with the valve-chests on top, steam distribution being by 10in dia piston valves driven by Walschaerts valve gear. The front of the running-plate was raised several inches to clear the valve-chests. The coupled wheels were 6ft in diameter, those of the leading and driving axles being enclosed in a common splasher. Front-end lubrication was by a Wakefield No 1 10-feed mechanical lubricator. The boiler-barrel was composed of three rings, the front and rear of which were 5ft 3in in external diameter and overlapped the ends of the intermediate ring on which the dome, with its sliding regulator of Drummond type, was situated. The original Cowlairs general arrangement drawing shows a 24-element superheater with long return loops providing a superheating surface of 448sq ft but, in fact, short return loops were provided, the surface area being 350sq ft. The Belpaire outer firebox was 8ft long, its roof being secured to that of the inner firebox by direct stays, of which the first four rows were flexible. The rear of the grate was level while the front of the sloping portion was hinged to facilitate cleaning of the fire. Before the engine entered service the drop grates were removed. Boiler feed was by two Davies and Metcalfe No 11 hot water injectors intended to take water from a Smith feed-water heater, though this feature was removed before the engines entered traffic. Working pressure was restricted to 160lb/sq in by a pair of 2½in dia Ross pop safety-valves. The first two engines were given massive built-up chimneys with wind-deflecting caps which brought the height above rail to 13ft 3¾in; since this was 4¾in over the Caledonian loading gauge, these chimneys had to be reduced in height. The remaining four engines were given single-piece chimneys of acceptable height. In all cases, the chimney liner was nicely tapered and terminated in a properly radiused bell-shaped

Page 159 (*upper*) Lambie 4-4-0 No 18 at Beattock in August 1902; (*lower*) 956 class 4-6-0 No 957 at Carlisle

Page 160 (upper) Drummond 4-4-0 No 66 as rebuilt with large boiler; *(lower)* River class 4-6-0 No 941

structure, and so was closer to the theoretical ideal than was the native St Rollox product.

The frames were 1⅛in thick but were less strongly braced than those of Pickersgill's own 4–6–0s, the first of which was by then on the drawing board. The main journals were 8½in in diameter and 10in long, only moderately sized by Pickersgill's standards. The proportion of reciprocating weight balanced must have been low, for the maximum hammerblow per axle at a speed of 6rev/sec was only 4.2 tons. Laminated springs were underhung below all six driving axleboxes but bogie suspension was by independent coil-springs of Timmis section. The brake-rigging was compensated throughout but its arrangement was unusual. There were two vacuum brake cylinders, each of 21in dia. One was situated between the leading and driving axles and applied blocks to the rear of the driving and trailing wheels while the other, below the footplate, was connected to blocks at the front of the leading and trailing wheels. Vacuum was provided by a No 25/20 Gresham Dreadnought combination ejector, but a Westinghouse pump, No 6 brake-valve and combination valve were fitted to enable the engine to work Westinghouse-fitted vehicles. Sanding (gravity) gear was supplied for forward running only while reverse was by a Drummond horizontal steam reverser between the frames. The six-wheel tenders had very high sides and were probably the heaviest of their type in the entire country.

The first two engines, which had been named *River Ness* and *River Spey* by the Highland Railway, were fitted with standard McIntosh numberplates; the other four engines were modified by the builders and got Drummond-pattern numberplates. In addition to those already mentioned, the alterations required by the Caledonian included the removal of the tender coal guards and the addition of supporting pillars to the cab roof. The Highland names were not retained. The engines were given express livery and were numbered 938-43 in order of construction. The builders' plates were on the smokebox side and bore the numbers 3095-3100.

Nos 938-41 went to Balornock and Nos 942-3 to Perth. The four Balornock engines seem to have been used on fast goods trains from the start but, until the end of 1916, the Perth pair regularly worked the 10am and 5pm expresses from Buchanan Street, usually week about, and on these duties are said to have proved at least the equal of a superheated Cardean. Thereafter they spent most of their time working fitted goods trains between Perth and Carlisle. Two of the Balornock engines were employed day about on the 7pm Buchanan Street (Goods) to Carlisle and the corresponding down working of the following day while the other two were similarly used for the 12.2am 'sweep' to Carlisle and its return working on the following day, the pairs changing round each fortnight. For such work they were superior to anything else owned by the company. Table 11 reproduces a guard's log of a typical war-time journey.

TABLE 11
RIVER CLASS WORKING

Date: 31/12/17
Engine No: 943

	Arrive	Depart	Wagons	Remarks
Kingmoor	—	6.05pm	48	
Quintinshill	6.30	6.50	48	Clean fire and take water
Beattock	7.55	8.05	48	Banked by No 441
Carstairs	9.30	9.50	48	Clean fire and take water
Carstairs No 1	9.55	10.00	48	
Law Jct	10.30	10.48	40	
Fullwood Jct	11.00	11.10	40	
Greenhill	?	11.45	40	
Larbert Jct	11.55	12.00	40	
Stirling	1.00am	1.14am	40	
Bridge of Allan	1.26	1.30	40	
Dunblane	1.38	1.40	40	
Greenloaning	2.13	2.23	40	
Perth	3.10	—		

Delays: Larbert (water) 12min
Plean (signals) 7min
Polmaise (signals) 10min

THE 4-4-0s

Construction of six superheated 4–4–0s to Pickersgill's designs had been authorised by the traffic committee on 6 October 1914 but, due to the shortage of labour, it was not until 1916 that the engines were ready:—

Order No	Engine Nos	Dates delivered	Cost per engine
Y113	113-6, 121/4	2/16—5/16	£2914

No 121 replaced the McIntosh engine of that number broken up after the Quintinshill disaster.

By the middle of 1915, the shortage of skilled workmen had become critical, and the company had some 4–4–0s built by private contractors. Quotations for 10 engines were considered on 13 July

1915 and the North British Locomotive Company's offer of £4235 per engine was accepted. A separate quotation for ten tenders— £7200—was rejected and the tenders built at St Rollox (Order No Y118). The engines were built at the Atlas Works in 1916 to order No L664 and were given the NBL works numbers 21442-51 and the Caledonian numbers 928-37.

The new 4-4-0 was a stouter version of the later McIntosh engine with a front-end based on that of the 179 class 4-6-0s and six-wheeled 4200-gallon tenders. The frames were increased to $1\frac{1}{4}$in in thickness and lengthened by 6in, bogie wheelbase being increased to 7ft and sideplay to $2\frac{3}{4}$in. Instead of the built-up dragbox of the McIntosh engines a heavy casting was provided. The cylinders, 20in in diameter, were brought 3in closer together and fitted with combined drain-cocks and spring-loaded pressure-relief valves. By-pass valves, which opened automatically when the regulator was closed, put the two ends of each cylinder into communication. The piston valves were standard but were increased to 9in dia. The motion was heavier than before, $\frac{1}{4}$in being added to the diameter of the crank-arms, journals, wheel seats and outside crank-pins, though the thickness of the inner crank-webs was *reduced* by $\frac{1}{4}$in. The repositioned cylinders allowed an increase in the length of the driving journals:—

		McIntosh	Pickersgill
Driving journals:	dia, in	$9\frac{1}{4}$	$9\frac{1}{2}$
	length, in	$7\frac{1}{2}$	9
Trailing journals:	dia, in	$9\frac{1}{4}$ & $7\frac{3}{4}$ (concave)	$8\frac{1}{4}$
	length, in	12	12
Bogie journals:	dia, in	6	$7\frac{1}{2}$
	length, in	11	10

Dimensions such as these ensured that the new engines would be capable of withstanding the overloading and abuse of wartime operating conditions.

The boiler was similar to that of the McIntosh 4-4-0s but the barrel contained six fewer tubes and the firebox was 3in shallower at the back. The brick arch was shorter and the superheater elements were of the short return loop variety, the loops extending only to the dome. Although these elements provided a lower degree of superheat than the long-looped type, the drop was not proportional to the reduction in surface area since, due to the temperature gradient between firebox and smokebox, it is the forward part of the superheater which does least work. Protection of the elements was given by the usual hand-operated damper. The smokebox arrangement was little changed though the blastpipe orifice was slightly lower and $\frac{1}{4}$in greater in diameter. The chimney, however, was a new and handsome one-piece casting. The St Rollox-built engines had two Ross pop safety valves housed in a conical casing while the NBL batch had four 4in lock-up valves in 903 class casings. Boiler feed was by two Gresham & Craven No 10 combination injectors in No 9 cases.

Batch Y113 and Nos 928-32 of batch Y118 were fitted with the company's own mechanical lubricator (No 2 pattern), worked from an auxiliary crank on the outside crank-pin of the driving axle. Nos 933-7 got Detroit 5-feed sight-feed lubricators instead. Sanding was by steam and involved the introduction of another new fitting in the cab, a combined steam sanding and blower valve attached to the faceplate to the left of the left-hand water-level gauge. The engines were dual fitted, vacuum being provided by a Gresham & Craven Dreadnought ejector. The brake cylinder was of larger diameter than that of the McIntosh engines and the Westinghouse pump was bolted to brackets rivetted to the firebox shell.

The new engines were distinctive in appearance though tradition was fully maintained. Smokebox wing-plates were dispensed with and the coupling rod splasher made straight from end to end. The NBL-built engines had Drummond numberplates and cone-shaped buffer casings. But the most noticeable change was the six-wheel tender holding 4200 gallons of water and six tons of coal. The tank was the same length as that of the Lambie-McIntosh 3570-gallon pattern but 9in wider and $4\frac{1}{2}$in deeper. The well, however, was slightly smaller and the concave journals, standard for so many years, gave place to cylindrical ones.

The 16 engines were allocated as follows:—

Carlisle	Nos 113-6, 121/4	Dalry Road	Nos 932-3
Polmadie	Nos 936-7	Balornock	Nos 934-5
Perth	Nos 928-31		

Although contemporary commentators have dismissed them as lifeless and sluggish, with neither the power nor the freedom of their predecessors, scrutiny of logs shows that they displayed commendable liveliness in uphill work and were by no means the only engines to run sluggishly downhill. A similar conclusion was drawn by O. S. Nock in *The Caledonian Dunalastairs*:—

A study of these runs, and particularly those with the Pickersgill engines, shows such consistent characteristics of good uphill work and slow running downhill as to suggest that a marked change had come over the whole attitude to express running on the Caledonian line.

The by-pass valves allowed them to run freely down the steeper gradients with the regulator shut and the restrained running downhill may sometimes have been due to that.

The valve-setting of the new 4–4–0s was based on that of McIntosh's 139 class. When Nos 928-37 were being built, a copy of the McIntosh valve-setting table was sent to the Atlas works, where it was stamped with the order No L664 and, in due course, with the later order No L768. The sole change made to the table was the addition of an instruction that 'for engines still to be delivered at this date 5/5/16 and in future', the valve setting was to be altered as follows:—

	Old setting		New setting	
	Front port	Back port	Front port	Back port
Forward gear				
Notch 5 from centre	$\frac{1}{2}$in	$\frac{1}{2}$in bare	$\frac{7}{16}$in	$\frac{9}{16}$in
Backward gear				
Notch 3 from centre	$\frac{3}{8}$in bare	$\frac{3}{8}$in bare	$\frac{3}{16}$in full	$\frac{5}{16}$in full

The smallness of the changes disposes of the oft repeated claim that the engines were handicapped by poor valve-setting.

THE 60 CLASS

Soon after his appointment, Pickersgill revived the plan for an outside-cylinder 4–6–0, though with 6ft 1in coupled wheels instead of the 6ft 6in wheels of McIntosh's abortive design. Construction of the first engine was sanctioned by the traffic committee on 6 October 1914 for the half-year ending 30 June 1915. Authority for five additional engines, to be built during the half-year ending 31 December 1915, was given on 9 March 1915. However, due to the shortage of labour, completion of the programme was considerably delayed:—

Order No	Engine Nos	Dates delivered	Cost per engine
Y115	60	11/16	£3711
Y116	61-5	12/16 – 4/17	£3948

The front-end was based on the McIntosh design. The 20in by 26in cylinders were cast complete with their valve chests and

TABLE 12
VARIATION IN LEAD WITH ADVANCING CUT-OFF
CRANK AT REAR DEAD-CENTRE. REVERSING LEVER MOVED FORWARDS.

Class	Movement of expansion link	Eccentric arrangement Diagram No	Movement of die-block	Rocker	Movement of valve	Admission	Lead
721	Lowered	1	Backwards	No	Backwards	Outside	Reduced
903 (slide-valves)	Lowered	2	Forwards	Yes	Backwards	Outside	Reduced
179	Lowered	1	Backwards	Yes	Forwards	Inside	Reduced
60	Lowered	2	Forwards	No	Forwards	Inside	Reduced
956 (Nos 957-8)	Raised	3	Forwards	Yes	Backwards	Inside	Increased

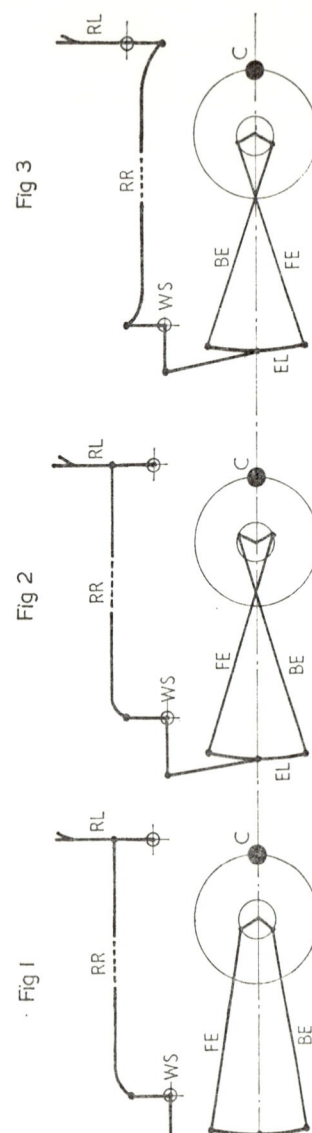

Fig 1   Fig 2   Fig 3

EL Expansion link   FE Fore-gear eccentric rod   BE Backward eccentric rod   RR Reach rod
RL Reversing lever   WS Weightshaft   C Crank

fitted with by-pass valves. Steam distribution was by 9in dia standard piston valves, the valve chests being between the frames, the most convenient position for the inside Stephenson link motion. Since the valve-spindles were driven direct, and admission was from the inside edges of the valve, the eccentric sheaves were keyed to the axle in the same way as those of the 49 class (slide valve condition), the eccentric rods therefore appearing crossed when viewed with the crank at its rear dead centre. This point deserves emphasis, for it has been claimed that the 'crossed' rods resulted in a decrease of lead as the engine was notched up and that this was responsible for the alleged sluggishness. This was not so; when the reversing lever was put into forward gear, with the big-end at the rear dead-centre, the expansion link was lowered but, because of the angularity of crossed rods, the die-block in the link was moved forward instead of backwards so that the valve was pushed forward and, admission being from the inner face of the valve-head, the port opening reduced. As the engine was notched up towards mid-gear, the movements took place in the opposite direction, lead therefore increasing. Study of Table 12 will clarify the position.

The frames and motion were built on a massive scale and, although official figures are not available, the experience of shed maintenance staff suggests that the new engines had greater freedom from mechanical troubles than their inside-cylinder forebears. In particular, wear of the axleboxes and axlebox-guides was much reduced. The frames were $1\frac{1}{4}$in thick and, in addition to the usual cross-stays, were braced by a pair of diagonally-arranged plates which provided great rigidity. All three coupled axles were interchangeable and had very large journals, of $9\frac{1}{2}$in dia and 11in long, providing 5 per cent greater bearing surface than those of the 903 class crank-axle. The crosshead was secured to the piston-rod by a cotter and ran between the channelled bearing surfaces of the two slide-bars. The connecting-rods were 11ft long, their little-ends being forked and their big-ends fitted with cotters and adjustable brasses. A spherical bearing in the joint uniting the two lengths of the coupling rod allowed the trailing-axle a side-play of $\frac{1}{4}$in. The expansion-links were hung behind the leading coupled axle, the eccentric-rods being 5ft 3in long. The valve-spindle was very long, its rear end being curved like an inverted U to clear the leading coupled axle. Tail-rods were fitted to the piston-valves as well as to the pistons. The bogie wheelbase was increased to 7ft and 3½in side-play provided in place of the usual 1½in. Laminated springs were used for all three coupled axles.

The boiler was a shortened version of that fitted to the 903

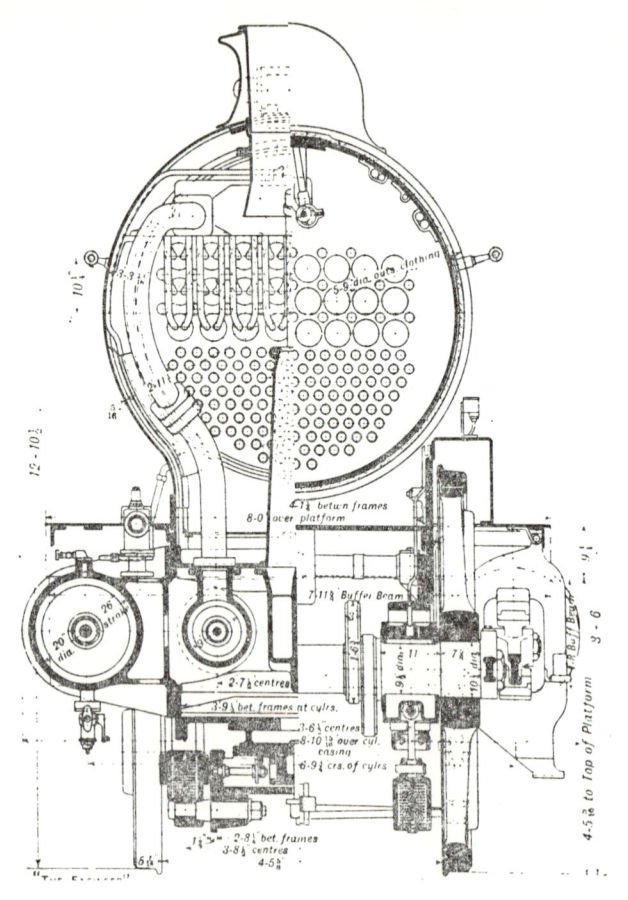

Fig 20   60 class end elevation

class, length between tubeplates being reduced by 1ft 5in; the barrel was telescopic and composed of three rings. The firebox was 2in deeper at the back than that of the 903 class but was otherwise unaltered. The brick-arch was longer and the rear of the ashpan, though fitted with a damper, was of simplified design without a hopper. The 24-element superheater had short return loops. A superheater damper was not provided, burning of the elements being prevented by two automatic snifting valves, housed in a single casing behind the chimney and communicating with the saturated side of the header. On closure of the regulator, these valves opened automatically and allowed air to be drawn through the elements. The smokebox was 1ft 4in longer than that of the 903 class, though the blastpipe and chimney were proportioned in the late McIntosh manner. No 60 was fitted with two hooded pop safety valves as used on the 113 class 4-4-0s while the remainder got four 4in lock-up valves.

The cab was approximately 5in shorter and 2in narrower than that of the 903 class, the lean-out being considerably smaller and the windows were of the type used in the 908 and 918 classes. Reverse was by steam-assisted lever, the reversing cylinders being mounted inside the frames ahead of the driving wheels. Other footplate fittings included a Gresham & Craven No 40mm vacuum ejector, two Gresham & Craven No 10 combination injectors, and a combined steam sanding and blower valve. A pyrometer was fitted to Nos 60 and 65. The main air reservoir was formed by the dragbox casting and the pump was bolted to brackets rivetted to the firebox shell. Front-end lubrication was by a Caledonian 10-feed mechanical lubricator mounted on the left-hand running plate and driven from an auxiliary crank on the trailing wheel. The six-wheel tenders were of the type fitted to the contemporary 4-4-0s.

The engines were allocated as follows:—

    Balornock   No 60   Carlisle   Nos 62-3
    Polmadie   No 61   Perth   Nos 64-5

No 60 ran between Buchanan Street and Perth. She was driven by 'Geordie' Mackie who, with trains of around 300 tons, frequently worked her half-way between notches 3 and 4 and claimed that nothing could hold her back. No 61 went to David Gibson and replaced *Cardean* on the Corridor. The Carlisle engines worked to Glasgow and Edinburgh and the Perth pair to Glasgow and Aberdeen. Trials run between Glasgow and Perth showed that the new engines could haul 370 tons at the average speed of 42mph demanded by the timetable for the best trains. It is recorded that outputs of 900-1000ihp were attained. In skilled hands, the class was capable of good work; fairly brisk running was possible on an early cut-off though this tended to cause heating of the eccentric-straps. Steaming was satisfactory provided a low water-level was maintained, certainly not higher than half-glass. However, on the southern main-line, the new engines never equalled, let alone surpassed, the finest achievements of the 49 and 903 classes in their saturated form and O. S. Nock has referred to them as 'curiously ineffective engines'. In *The Engineer* of 15 October 1920, E. C. Poultney described a run made on the footplate of No 61, working the 3.50pm Liverpool and Manchester express from Glasgow Central, and this certainly bears out Mr Nock's contention. Even allowing for the heavy load of 389 tons tare, the time of 26min 10sec for the 17.2 miles from Symington to Beattock Summit, start to pass, compares unfavourably with the work done by No 49 in her heyday. Although the train had left Carstairs 16¾min late, down hill speeds were restrained and only ¼min had been regained by Carlisle.

Sluggish or restrained downhill running was a hallmark of 60 class performance and commentators have tended to blame the valve-setting. However, the fact that speeds of up to 65mph could be attained on an early cut-off must surely exonerate the valve-setting.

CHAPTER 11

WAR AND PEACE

NEW GOODS ENGINES

Towards the end of 1917, the 25 requisitioned 'Jumbos' left Scotland for service in Europe. Their numbers were:—

Batch	Engine Nos	Batch	Engine Nos
Neilson	294, 517, 680/2	SW	365 (ex 695)
Y5	310/5/8-9	Y9	367
Y12	323 (ex 339)	Y23	403
Y29	374	Y31	548-9, 553
Y32	558	Y37	259, 260, 335/7
Y38	703/5/7-8		

Although the company had received a year's warning of their requisition, it seems to have had doubts about their eventual return, for the locomotive register recorded them as 'Sold to Government. Broken up 31/12/17'. Their replacement was considered by the traffic committee as early as 5 September 1916:—

> Submitted memorandum from the General Manager in regard to the building of new engines. It will be necessary to build 25 six-coupled Tender Goods Engines during the year 1917 under renewal account and as it will be impossible for more than 10 of these to be built at St Rollox, it is recommended that the remaining 15 should be built by outside builders and that tenders should be obtained.
> Approve taking tenders.

However, when the matter was discussed by the locomotive and stores committee on 3 October 1916, it was agreed that the general manager should seek an agreement with the Railway Executive Committee for the delivery of material for the repair of existing engines rather than proceed with the purchase of new engines.

171

WAR AND PEACE

The drawing-office next embarked on the design of a heavy 2-6-0 with front-end and motion based on that of the 60 class but with 21in by 28in cylinders, 5ft 8in coupled wheels and a larger boiler and firebox. The general arrangement drawing, which was dated 26 July 1917 and signed by Pickersgill, shows several unusual features. The 24 large flues were only $3\frac{1}{4}$in in diameter and the return loops of the superheater elements were omitted altogether, superheating surface being no more than 165sq ft. Another novel feature for a St Rollox-designed engine was that the brakes were powered by vacuum. However, the shortage of skilled workmen and escalating costs, put paid to this bold project.

McIntosh's 30 class was taken as the basis for a new class of 0-6-0s. The front overhang was shortened by 1ft 9in and the tail-rods abandoned. The cylinders were reduced to $18\frac{1}{2}$in dia and the piston valves enlarged to 9in dia. Front-end lubrication was by a Caledonian No 4 mechanical lubricator worked by an auxiliary crank from the outside crank-pin of the driving axle. The superheater was omitted altogether and the inner firebox of the first twelve engines constructed of steel instead of copper. (New copper fireboxes were fitted between January 1920 and October 1922.) The smokebox was shortened by 5in and surmounted by a standard Pickersgill chimney. The frames were increased to $1\frac{1}{4}$in in thickness and spiral springs used for the driving axleboxes. The dragbox was redesigned and steam sanding gear fitted to both ends of the engine. The tender was of the McIntosh 3000-gallon pattern with coal-rails added. All engines were steam-braked and painted black.

The McIntosh 30 class 0-6-0s had cost £2312 each, but inflation greatly increased the cost of the new 0-6-0s.

Order No	Engine Nos	Dates delivered	Cost per engine
Y119	300-9	2/18 – 6/18	£3480
Y120	310-1		£3480
	312-5	7/18 – 5/19	£4499
	316-21		£4485
Y122	294-9, 322-4	6/19 – 9/19	£4531

At the end of 1919, 12 engines of modified design were turned out:—

Order No	Engine Nos	Dates delivered	Cost per engine
Y123	280-1, 670-9	10/19 – 1/20	£4597

TABLE 13
Y123 (PICKERSGILL 0-6-0) VALVE SETTING

Notch	Travel, in	Lead, in FP	Lead, in BP	Port opening, in FP	Port opening, in BP	Cut off, % FP	Cut off, % BP	Release, % FP	Release, % BP
9	$4\frac{7}{8}$	0	$\frac{3}{16}$	$1\frac{1}{2}$B	$1\frac{3}{8}$	$86\frac{1}{2}$	82	96	95
8	$4\frac{1}{2}$	$\frac{1}{32}$	$\frac{7}{32}$	$1\frac{1}{4}$	$1\frac{1}{4}$	$83\frac{1}{2}$	$78\frac{1}{2}$	94	94
7	$4\frac{1}{8}$	$\frac{1}{16}$	$\frac{1}{4}$	$1\frac{1}{16}$	$1\frac{1}{16}$	80	73	93	92
6	$3\frac{3}{4}$	$\frac{1}{16}$F	$\frac{5}{16}$B	$\frac{7}{8}$	$\frac{7}{8}$	76	68	91	90
5	$3\frac{7}{16}$	$\frac{3}{32}$	$\frac{5}{16}$	$\frac{11}{16}$	$\frac{3}{4}$	70	61	89	87
4	$3\frac{1}{8}$	$\frac{1}{8}$B	$\frac{3}{8}$B	$\frac{1}{2}$	$\frac{5}{8}$	62	50	85	82
3	$2\frac{7}{8}$	$\frac{1}{8}$	$\frac{3}{8}$	$\frac{3}{8}$	$\frac{1}{2}$	56	40	81	76
2	$2\frac{11}{16}$	$\frac{1}{8}$F	$\frac{3}{8}$F	$\frac{1}{4}$	$\frac{7}{16}$	41	31	76	69
1	$2\frac{5}{8}$	$\frac{5}{32}$	$\frac{3}{8}$	$\frac{3}{16}$	$\frac{7}{16}$	30	22	70	60

Lap of valve 1in
F=full B=bare

In these, the inner firebox was made of copper while the piston valves were replaced by slide-valves balanced by a large ring sprung into the back of the valve (the valves lay horizontally above the cylinders). This had been a familiar feature of Inverurie practice during and before Pickersgill's period there. The tender springs of this batch had adjustable hangers through pads rivetted to the frames.

The Pickersgill 0-6-0s were concentrated in the Forth-Clyde valley though Aberdeen got three, Forfar two and Perth four.

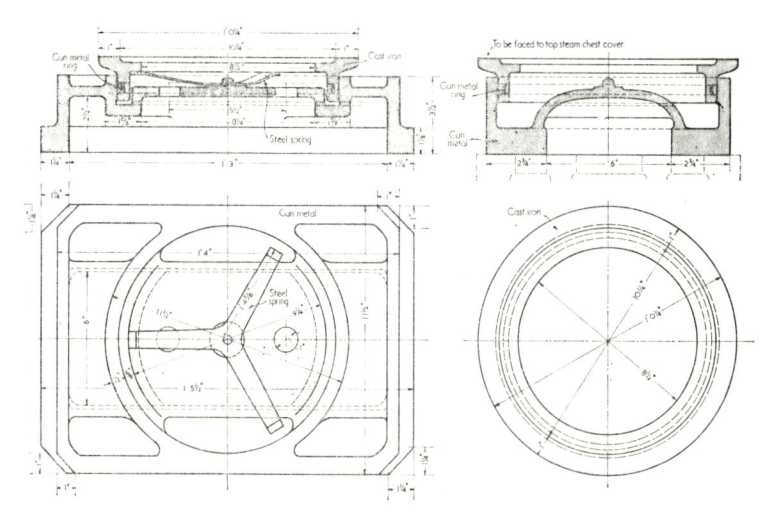

Fig 21 Pickersgill balanced slide valve

Otherwise, the northern and southern limits of their allocation were Stirling, with six engines, and Carstairs, also with six.

TANK ENGINES: 1915-22

Only one new class of tank engine was built during the remainder of the company's existence. This was basically a scaled-down 60 class 4–6–0 with the 4–6–2 wheel arrangement. The cylinders were reduced to 19½in dia and inclined more steeply to suit the 5ft 9in coupled wheels. Bogie wheelbase remained at 7ft but the coupled wheelbase was reduced by 1ft 3in. The valve gear was modelled on that of the tender engines but with the intermediate valve spindle and eccentric-rods reduced in length to suit the shorter wheelbase. The axles were interchangeable with those of the tender engines. Reverse was by the standard steam-assisted lever, the reversing cylinders being attached to the frame between the leading and driving coupled wheels. The rear of the engine was carried on a pony truck whose 3ft 6in dia wheels had a total side-play of 5in. Clearance was provided by cranking in the frames from a point just to the rear of the trailing coupled wheels. Laminated springs were used for all axles.

The boiler-barrel consisted of three rings, the front and rear of which were 4ft 9¼in in external diameter and overlapped the ends of the intermediate ring. With an outer firebox only 7ft long and with water legs 2½in wide, grate area was restricted to 21.5sq ft. The rear half of the grate was level to provide the maximum depth at the back of the firebox. The roof of the copper box was secured to the outer firebox casing by direct stays. The 21-element superheater was of the standard type with short return loops. Working pressure was limited to 170lb/sq in by two 2½in dia hooded pop safety valves. Although the Westinghouse pump was situated in front of the right-hand side-tank, the opportunity of extending the cab to the full width of the tanks—8ft 6¾in—was not taken; the 7ft wide cab proved rather cramped for it not only had to accommodate the reversing lever but also the mechanical lubricator, which was of the Wakefield No 7A pattern and was mounted against the side-sheeting on the fireman's side, the drive being from the crank-pin of the trailing coupled wheel. Other cab fittings included steam heating apparatus and a vacuum ejector and combination valve. Steam sanding gear was fitted for backward as well as for forward running.

WAR AND PEACE

Twelve engines were authorised on 16 November 1915. Fourteen days later, the locomotive and stores committee accepted the North British Locomotive Company's tender of £5250 per engine, the 12 to be delivered within nine months. The actual delivery dates ranged from March 1917 to May of the same year. The engines were painted blue and numbered 944-55, the numberplates being of the non-standard type used on the 92 class. Nos 944-5 went to Polmadie and for a time worked stopping trains between Glasgow and Edinburgh. Nos 951-3 and 955 went to Beattock for banking duties. They were not popular there and, after about five years, were transferred to Polmadie. No 950 was also at Beattock for a short time in 1917. With a greater appetite for coal than the 0–4–4Ts but with a bunker capacity only 10cwt greater, they required coaling more frequently; coaling facilities at Beattock were primitive, small tubs of coal having to be hoisted by hand-crane and tipped manually. Another difficulty was that leakage occurred from the ends of the large flues, no doubt due to the drop in smokebox temperature as the engine coasted home after a hard climb. No 954 was sent to Dawsholm for working heavy football specials over the low-level lines. The remainder went to Greenock for the Wemyss Bay and Gourock services, their association with the former branch earning them the nickname 'Wemyss Bay tanks'. Water capacity was 1800 gallons and coal 3 tons.

During the Pickersgill period, the 498 class 0–6–0Ts were expanded in number, three batches being constructed, the cost of the engines rising as the war progressed:—

Order No	Engine Nos	Dates delivered	Cost per engine
Y111	527-32	4/15–6/15	£1136
Y121	533-8	8/18–10/18	£2292
Y126	502-4, 510-5	/20– /21	

These engines followed the original McIntosh design. Several went to Edinburgh, one to Leith. Balornock got one, while the remainder were divided between Polmadie, Greenock and Dawsholm, the latter boarding out one or two of its engines to its sub-shed at Yoker.

Two further batches of standard inside-cylinder 0–6–0Ts were constructed:—

Order No	Engine Nos	Dates delivered	Cost per engine
Y117	232-5, 272-4	6/16–9/16	£1682
Y127	248-55, 394-6	/21– /22	

WAR AND PEACE

Those of batch Y127 were initially placed in store, No 397 being at Dawsholm for over seven months before use.

Pickersgill was also responsible for two batches of standard 0-4-4Ts:—

Order No	Engine Nos	Dates delivered	Cost per engine
Y114	159, 161-3	8/15—10/15	£1602
Y130	418, 425-6, 430, 435-6 431-4		

In these engines the bogie and engine wheelbase were lengthened by 6in and steam sanding gear fitted, the trailing sandboxes being placed under the running-plate, behind the driving wheels. The axleboxes and axlebox guides were modified and the bogie journals were of increased dimensions. Nos 431-4 were specially designed for banking duties at Beattock and had cast-iron front buffer-beams—which increased their length over buffers by 4in—18¼in by 26in cylinders and a pair of uncased Ross pop safety valves allowing a working pressure of 180lb/sq in. The remaining engines of batch Y130 also had pop safety valves though set to 160lb/sq in.

THE LATTER-DAY SCENE

By the middle of 1917, the wartime disruption was at its height. When a correspondent of *The Railway Magazine* visited Buchanan Street station he saw a recently arrived express from Perth headed by a 0-6-0 while, at the adjacent platform, stood the almost new No 60 at the head of a Glenboig-Glasgow local. The 6.10pm express, consisting of five 38-ton 12-wheelers, went out behind 'Coast Bogie' No 1115 while 4-6-0 No 50 was on a workmen's train of two ancient vehicles.

The Caledonian glamour had worn thin.

The company struggled towards the Armistice as best as it could, but much of the sparkle had gone. With the exception of *Breadalbane*, all locomotive names were removed. The loss of *Dunalastair* and *Cardean* was a sign of the times.

Many of the more famous non-superheated McIntosh engines were relegated to secondary services. Nos 723, 774 and 779 were shedded at Stirling for a time, 779 being sub-shedded at Callander for working the morning business train to Glasgow and the corresponding return working in the late afternoon. Beattock got No 778 for the Tinto residential express. By the grouping, about half a dozen of the Dunalastair Is were at Greenock and one, No 725, was at Adrossan. The Dunalastair II class was largely distributed between Polmadie and Balornock, which also got four of the 900s, while Carstairs acquired three of the 140s, Nos 925-7. With the introduction of the 4-6-2Ts on the Gourock line, five of the 908 class, Nos 908-12, were transferred to Balornock where they worked out the rest of their lives on goods trains to Carlisle and Perth.

Another noticeable change occurred in 1919 with the standardisation of Ross pop safety valves for all new boilers. Commencing with No 204 in 1917 the condensing apparatus was removed from the 1, 19, 29 and 92 classes and, by the Grouping, most if not all of the 1 and 29 classes had been so treated. Another development about this time was the introduction of tender cabs, these being fitted to 'Jumbos' Nos 594, 596 and 744, Dunalastair No 726 and 'Coast Bogies' Nos 195/7-8. These three engines were sent to Stranraer, their tender cabs probably being intended to provide protection on the exposed climb out of Portpatrick. Steam sanding equipment was sometimes fitted to rebuilt engines or to those receiving new boilers or frames. Steam heating and vacuum brake equipment were fitted to some of the smaller passenger engines from about 1920 onwards. The larger engines were little altered in appearance but *Cardean*, No 905, acquired hooded pop safety valves and a large-diameter air-cooled Westinghouse pump.

Despite experiment with various types of lubricator and a change in the drive of the mechanical lubricators in the McIntosh engines (from crosshead to rocking lever so as to vary with cut-off), carbonisation of oil in the valve-chests and steam passages of the superheated engines remained a perennial problem. In a paper read to the Institution of Locomotive Engineers in March 1921, Irvine Kempt Junior defined the problem:—

> The trouble which is experienced in all engines of this class is due to the collection of burnt oil on the pistons and piston valves, which prevents the spring rings from working and so causes leakage of steam. . . . We have carbonisation in the cylinders of all superheater engines, no matter what lubricators are in use.

One of the devices which had been tried was the Robinson Intensifore forced-feed, sight-feed lubricator. This was fitted to the former

140 class engine No 923 which was rebuilt with a 24-element superheater, 20¼in by 26in cylinders and 8in piston valves in 1917. No 924, similarly rebuilt in May 1915, had received a mechanical lubricator, and when two further engines of the 900 class, Nos 894 and 900, were superheated in April 1916 and May 1918 respectively, mechanical lubricators were again fitted. The number of 4-4-0s rebuilt with superheaters since McIntosh's retirement averaged only one per annum and, after 1918, no further conversions were carried out until December 1922 when No 147 was fitted with a 24-element superheater and a Detroit hydrostatic lubricator but was allowed to retain her slide valves and 19in by 26in cylinders. Thus modified, she was found to be a great improvement on the 140s generally. She often ran the 1.30pm Aberdeen express from Buchanan Street to Perth but took her turn on most of the Balornock duties, including those normally worked by 4-6-0s. The famous Geordie Mackie got her on the 5pm Aberdeen express one night when No 60 was off and said afterwards that she was the best 4-4-0 he had been on for years. The Detroit hydrostatic lubricator was fitted also to the saturated No 137.

In January 1923, just five months before the Caledonian ceased to exist as an independent company, 140 class No 927, which had been fitted with a Davies and Metcalfe exhaust steam injector in 1920, was tested on the 11.16am from Carstairs, her home shed, to Carlisle and the 6.32pm from Carlisle home. Three return trips were made, the exhaust steam injector being used on two and the live steam injector alone on the third. The results were as follows:—

Date of test		15/1/23	18/1/23	19/1/23
Injector		Exhaust	Live	Exhaust
Weather		High wind	Fine	Stormy
Load tons tare:	up	156	162	187
	down	201	205	206
Average speed, mph:	up	46.4*	45.9*	45.9
	down	44.1	44.5	44.1
Water consumption: total gallons		5603	5560	6095
Total coal used, cwt		70	75	82
Coal per train-mile, lb		53	57	62½

* Inclusive of stop at Symington

In the down direction the usual stop was made at Beattock for banking assistance. Although inconclusive, the results did not suggest that any marked economy was likely to result from the fitting of exhaust steam injectors. In view of the moderate loads, the coal consumption of over 50lb per mile was high but it must be remembered that the engine was 15 years old and no longer a top-link engine.

In 1919-20 the Caledonian borrowed standard Great Central Railway 2–8–0s, built for war service with the Railway Operating Division (ROD) but never so used. At least 53 of the class are known to have run on Caledonian metals, their ROD numbers being 1615, 1701-3, 1705, 1710-2, 1745, 1808, 1812, 1815-8. 1824, 1834, 1838, 1846, 1994, 2077-84, 2086-7, 2125-37 and 2158-67. The company's initials were painted on the tender, with a full stop between them, but no further alterations seem to have been made. The engines were distributed between Dawsholm, Balornock and Aberdeen and were returned to the government about the end of 1921. A G2 class 0–8–0 No 2430 was borrowed from the LNWR and, together with a ROD engine, put through a series of tests between Motherwell and Craigenhill.

After the war, 32 Pickersgill 4–4–0s were added to stock. Nos 72-81 were constructed at St Rollox to order No Y124 between May and September 1920. Nos 80-1 had Menno grease lubricators on the eccentric straps. Armstrong, Whitworth and Co supplied Nos 82-91 (works numbers 111-20) between February and June 1921 while the Hyde Park works of the North British Locomotive Company completed Nos 66-71 and 92-7 in November and December 1922 (works numbers 22943-54). These engines differed from the 1916 engines in having $20\frac{1}{2}$in by 26in cylinders, coil springs for the driving axleboxes and uncased Ross pop safety valves set at 180lb/sq in. The snifting valve was situated behind the chimney and communicated direct with the header. The Armstrong, Whitworth engines were painted a much darker blue than was then customary and had non-standard numberplates of the type fitted to the 4–6–2Ts.

Thomas Weir retired in April 1920. Although he was not entitled to a pension, the board granted him £165 per annum plus an honorarium of £250 'in recognition of his long and valuable services'. He was succeeded by W. H. Moodie who had served in the drawing office since 1896 and was only three years his junior.

THE LAST GENERATION

Soon after the end of the war, the drawing office began work on a very large three-cylinder 4-6-0 for passenger traffic, and four engines were constructed at St Rollox in 1921:—

Order No	Engine Nos	Dates delivered	Cost per engine
Y125	956-9	6/21—8/21	

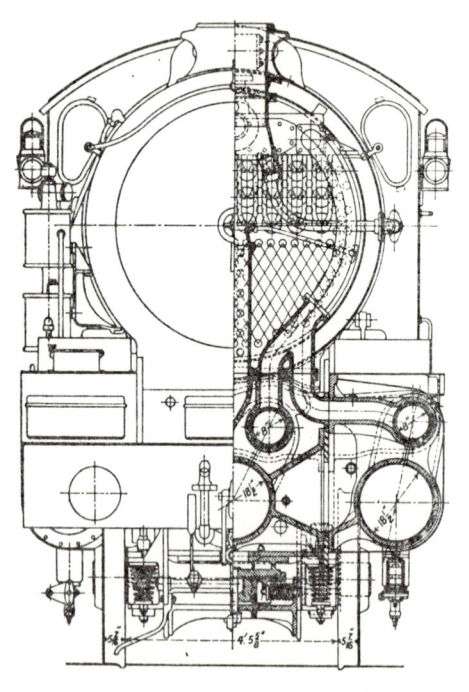

Fig 22 956 class end elevation

The 18½in by 26in cylinders were equipped with by-pass valves and arranged in line abreast, with the valve chests above them. The outside cylinders were inclined at 1 in 69 and drove the middle pair of coupled wheels through connecting rods 11ft long. The inside cylinder was inclined at 1 in 41 and drove the leading coupled axle through a connecting rod only 6ft 6in long. Each cylinder, together with its steamchest and pipes, formed a separate casting. The outside connecting rods had big-ends with cotters and adjustable brasses while the inside rod had the usual marine-type

TABLE 14
956 CLASS VALVE SETTING: FORWARD GEAR

Notch	Travel, in	Port opening, in FP	Port opening, in BP	Cut off, % FP	Cut off, % BP
10	$5\frac{7}{8}$	$1\frac{3}{8}$	$1\frac{1}{2}$	72	70
9	$5\frac{9}{16}$	$1\frac{7}{32}$	$1\frac{5}{16}$	68	66
8	$5\frac{3}{16}$	$1\frac{1}{16}$	$1\frac{1}{8}$	64	61
7	$4\frac{7}{8}$	$\frac{7}{8}$	$\frac{15}{16}$	59	56
6	$4\frac{9}{16}$	$\frac{3}{4}$	$\frac{25}{32}$	53	51
5	$4\frac{9}{32}$	$\frac{19}{32}$	$\frac{5}{8}$	$45\frac{1}{2}$	43
4	4	$\frac{15}{32}$	$\frac{1}{2}$	37	36
3	$3\frac{3}{16}$	$\frac{3}{8}$	$\frac{3}{8}$	$27\frac{1}{2}$	27
2	$3\frac{5}{8}$	$\frac{5}{16}$	$\frac{5}{16}$	19	19
1	$3\frac{17}{32}$	$\frac{1}{4}$	$\frac{1}{4}$	13	15

Lap of valve $= 1\frac{1}{8}$ in   Lead $= \frac{1}{4}$ in

big-end. Coupled wheels and journals were of the same size as those of the 60 class. Reverse was by steam-assisted lever, the reversing engine being in the cab.

Steam distribution was by 8in dia piston valves. Those serving the outside cylinders were driven by Walschaerts valve gear but the inside valve was driven by a complex derived motion designed in the autumn of 1919, and illustrated in the accompanying diagram.

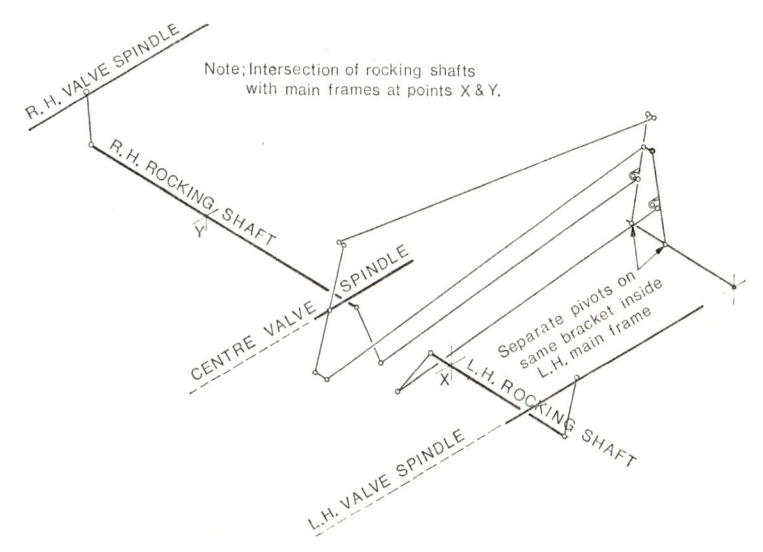

Fig 23 956 class: arrangement of derived valve gear

The very large boiler barrel was composed of the usual three rings, the largest having an outside diameter of 5ft 9in. The outer firebox casing was 9ft 3in long and the evenly sloping grate provided an area of 28sq ft. Flanging blocks had to be hired from the North British Locomotive Company for the construction of these fireboxes as St Rollox had none of suitable size. The firebox crown

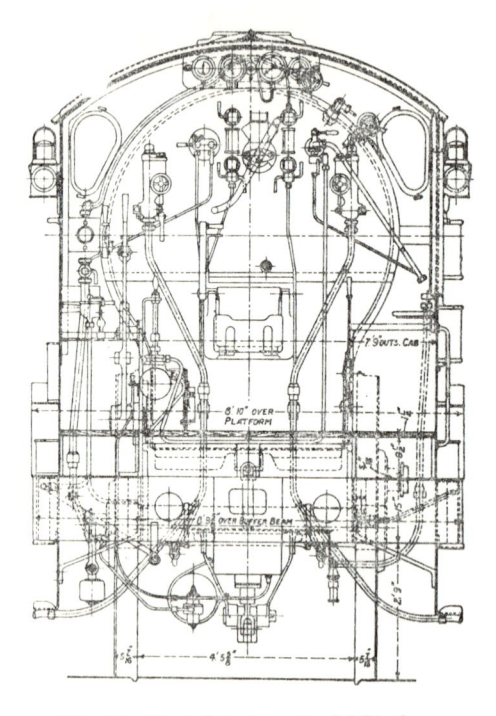

Fig 24 Footplate layout of 956 class

was direct-stayed while the 24-element superheater was of the standard type with short return loops. Protection of the elements was afforded by the snifting valves at the rear of the chimney. The Ross pop safety valves were set to blow off at 180lb/sq in.

Front end lubrication was by a Caledonian eight-feed mechanical lubricator, this also supplying the leading and intermediate coupled axleboxes. The bogie suspension was based on that of the River class, two independent spiral springs being fitted to each of the bogie axleboxes. The trailing coupled wheels had laminated

WAR AND PEACE

springs but the other coupled axleboxes were fitted with double coil springs. Steam sanding gear was fitted for forward running, the necessary valve being combined with the blower valve in Pickersgill fashion. The platform and splasher arrangement was of the River type and the cab roof was fully radiused. The engines were dual-fitted, vacuum being provided by a Dreadnought ejector. The cast-iron dragbox formed the main reservoir for the Westinghouse brake. The six-wheel tender was a widened version of the 4200-gallon pattern and could be distinguished by the less acute angle of its coal flare. Water capacity was 4500 gallons and coal capacity 6 tons.

TABLE 15
956 CLASS DIAGRAMS

Diagram No.	6	7	9	10	18
Incline (up-grade)	1 in 300	1 in 100	1 in 528	Level	1 in 75
Boiler pressure, lb/sq in	170	170	175	175	170
Regulator opening	$\frac{7}{16}$	Full	$\frac{1}{4}$	$\frac{1}{8}$	Full
Steam chest pressure, lb/sq in	150	150	140	125	165
Cut off, %	27	27	27	35	40
Mean effective pressure, lb/sq in	47	61	48	36	92
Speed, mph	52	43	57	62	23
Indicated horsepower	1194	1263	1335	1082	1053

No 956 went to Balornock, Nos 957-8 to Carlisle and No 959 to Perth. With indicator shelters fitted, No 956 ran trials in August 1921, her regular driver—James Grassie—taking her to Carlisle as well as to Perth. The indicator diagrams taken do not state which cylinder, inside or outside, was indicated but since the total horsepower was given as three times the horsepower of the indicated cylinder, it would seem that little difference was found between them. Five diagrams taken on 11 August, when the engine worked the 10am from Glasgow Central and the 3.35pm return working, are shown in Table 15. The maximum indicated horsepower recorded was 1335 but a clearer idea of the capacity of the locomotive is shown by Table 16 which compares the power developed with that of Maunsell's *King Arthur* at similar speeds and cut-off. These figures suggest that the engine was not the 'monumental failure' that latter-day writers would have it. Further tests were carried out in 1922; coming north with the Corridor, some 420 tons gross, No 956 is reputed to have reached Beattock in 45¾min.

Despite their potential, the three-cylinder 4–6–0s were disappointing. The derived valve gear was a source of trouble, though when it was new, it seemed to give satisfactory results. Indeed, the long-travel valves provided a greater port opening for any given cut-off than those of the Gresley A1 class Pacifics. (At a cut-off of 19 per cent, the port openings of the Caledonian engine were similar to those of a Gresley Pacific working at 25 per cent). Ultimately, wear of the pins resulted in malfunction of the inside valve. On some occasions bad steaming seems to have been the cause of poor performance and, on the faster stretches, riding tended to be rather wild, the extensive use of coil springs perhaps being responsible for this.

TABLE 16
HORSEPOWER COMPARISONS

Engine	Cut off, %	Regulator opening	Speed (mph)	IHP
956	27	Full	43	1263
King Arthur	$27\frac{1}{2}$	Full	41	896
956	27	$\frac{1}{4}$	57	1335
King Arthur	25	Full	57	1233
956	27	$\frac{7}{16}$	52	1194
King Arthur	$22\frac{1}{2}$	Full	51	1047

Early in their careers, probably in April 1922, Nos 957-8 were fitted with Stephenson link motion for the inside valve. It has been said that this was a blunder, because the variable lead of the Stephenson link motion was incompatible with the fixed lead of the outside Walschaerts gear. The real trouble was that the inside motion was arranged so that the lead decreased as the engine was notched up (see Table 12). At low speed with an advanced cut-off, the lead was excessive and the engine's powers of acceleration were adversely affected. At high speed on an early cut-off, it was probably insufficient for optimum performance. In 1922, No 959 was fitted with a modified form of the derived valve gear but she too was rebuilt with inside Stephenson link motion.

The temporary transfer of three 918s to the Oban line in the summer of 1914 suggests that the 55 class 4–6–0s were by then no longer in complete mastery of the heaviest trains. After the war, Pickersgill had more pressing problems to solve and it was not until the Grouping was imminent that a new class of 4–6–0s was built for the Oban line, eight engines—Nos 191-8—emerging from the Queens Park Works of the North British Locomotive Company in 1922. Carrying the works numbers 22955-62 and costing £5210 each, they were attractive little engines, better proportioned than the 55s and, theoretically, well suited to replace the ageing 4-4-0s on the more humdrum main-line duties.

TABLE 17
191 CLASS VALVE SETTING: FORWARD GEAR

Notch	Travel, in	Port opening, in FP	BP	Cut off, % FP	BP	Release, % FP	BP
7	$5\frac{1}{2}$	$1\frac{1}{2}$	$1\frac{1}{2}$	76	74	92	92
6		$1\frac{13}{64}$	$1\frac{13}{64}$	70	68	90	89
5		$\frac{7}{8}$	$\frac{7}{8}$	60	58	86	85
4		$\frac{21}{32}$	$\frac{21}{32}$	50	48	82	81
3		$\frac{1}{2}$	$\frac{1}{2}$	40	39	77	76
2		$\frac{3}{8}$	$\frac{3}{8}$	30	29	71	70
1		$\frac{9}{32}$	$\frac{9}{32}$	20	19	64	63

Lap of valve = $1\frac{1}{4}$in   Lead = $\frac{1}{4}$in

The $19\frac{1}{2}$in by 26in outside cylinders were inclined at 1 in 26, with the valve chests on top, and drove 5ft 6in dia coupled wheels. Steam distribution was by slide valves, balanced by a ring sprung against the top of the valve as in the last 0-6-0s. The valves were driven by Walschaerts gear arranged for outside admission and had a lap of $1\frac{1}{4}$in and a lead of $\frac{1}{4}$in. Travel in full gear was $5\frac{1}{2}$in. Reverse was by steam-assisted lever, the reversing cylinders being attached to the inside of the frame, opposite the expansion links. The frames were $1\frac{1}{8}$in thick and were diagonally braced as in the other Pickersgill 4-6-0s.

The boiler-barrel was composed of the usual three telescopic rings, the largest being 4ft $9\frac{1}{4}$in in external diameter. There was no superheater. The firebox crown was direct-stayed while the grate had a uniform slope and the ashpan was fitted with dampers at front and rear. The 3in dia Ross pop safety valves were set to blow off at 185lb/sq in. Boiler feed was by two Gresham & Craven No 10 combination injectors in No 9 cases. The bogie suspension was modelled on that of the 956 class but laminated springs were used for the leading and trailing coupled axleboxes. Steam sanding gear was fitted for forward running only. Front-end lubrication was by a Caledonian No 5 8-feed mechanical lubricator which also supplied the leading and driving coupled axleboxes. The runningplate and splashers were arranged as in the 956 and River classes.

The tender frames were similar to those of the 55 class but the tank was 6in shorter, 4½in less deep and 11¾in wider. Two coal-rails were fitted on either side. The axleboxes had a side-play of ¼in.

Nos 191-2 and 197-8 went to Balornock and were employed on local trains out of Buchanan Street but the remainder were put on the Oban line where they were later joined by Nos 191-2. They should have proved more versatile engines than the 55s but, although some of the men thought highly of them, the majority had a poor opinion of their capabilities and preferred the older engines. One mechanical problem which beset their early running was fracture of the valve-spindles.

Pickersgill's last project was for a heavy 2–6–0 but, by the time preliminary drawings had been prepared the Caledonian had become part of the LMS and the project was cancelled. With horizontal outside cylinders, 21in by 28in, outside Walschaerts valve gear and a shortened version of the 956 class boiler, it would have been a handsome and powerful locomotive.

Along with the North Staffordshire Railway and several minor concerns, the Caledonian entered the LMS on 1 July 1923 and therefore enjoyed six months of independence denied to the other constituents. But it was an empty gesture for the end was nigh.

APPENDIX 1

CASE HISTORIES

Eng No	Reno'd	Reb'lt	LMS Nos	Eng No	Reno'd	Reb'lt	LMS Nos
	Class 123			68	1068	1903	14291
123	1123	—	14010	69	1069	1901	14292
						1911	
	Class 80			70	1070	1901	14299
				71	1071	1898	14300
	Order No Y13			72	1072	1891	14293
80-1	1081-1	1905	14108-9			1904	
82	1082	1904	14110	73	1073	1902	14301
85	—	1902	Wdn 1916	74	1074	1904	14294
86	1086	1904	Wdn 8/21	75	1075	1901	14302
116	1116	1904	Wdn 4/21		*Order No Y21*		
	Order No Y28			76	—	1905	Wdn 8/22
114	1114	1908	14111	77	—	1907	Wdn 8/22
115	1115 85	1904	Wdn 8/22	78	—	1904	Wdn 8/21
195-7	1195-7	1907	14112-4	79	1079	1899	14297
198	1198	1908	14115	84	1084	1906	14306
				87	—	1904	Wdn 1917
	Class 66				*Order No Y25*		
	Order No Y			83	1083	1909	14305
60	17 1060	1901	14303	88	1088	1907	Wdn 3/22
61	1061	1896	14295	89	1089	1899	Wdn 2/21
62	1062 87	1902	14304	90	1090	1906	14307
63	—	1902	Wdn 1916	91	1091	1902	Wdn 12/22
64	1064	1901	Wdn 12/20	113	1113	1904	Wdn 2/21
65	1065	1901	Wdn 4/18	124	1124	1906	14296
	Neilson				**Class 13**		
66	1066	1901	14298		*Order No Y35*		
67	1067	1903	14290				
		1915		13	1013	1903	14308

187

APPENDIX 1

Eng No	Reno'd	Reb'lt	LMS Nos
14	—	1905	Wdn 11/20
15	1015	1904	14309
16	—	1904	Wdn 10/22
17	—	1905	Wdn 1916
18	1018	1917	14310

Class 721
Order No Y44

721-31	—	—	14311-21
732	—	1908	14322
733-5	—	—	14323-5

Class 766
Order No Y51

766	—	1914	14430
767-8	—	1920	14326-7
769	—	1914	14431
770	—	—	14328
771-2	—	1914	14432-3
773-4	—	—	14329-30
775	—	1920	14331
776-7	—	—	14332-3
778	—	1920	14334
779-80	—	—	14335-6

Class 900
Order No Y57

900	—	1918	14435
901	—	1914	14436
902	—	1921	14346

Order No Y62

887	—	—	14337
888	—	1922	14338
889	—	—	14339
890	—	1919	14340
891	—	—	14341
892	—	1922	14342
893	—	1911	14343
894	—	1916	14434
895	—	—	14344
896	—	1920	14345
897	—	—	14347
898	—	—	14437
899	—	—	14348

Class 140
Order No Y72

140	—	1919	14349

Eng No	Reno'd	Reb'lt	LMS Nos
141-2	—	—	14350-1
143	—	1920	14352
144	—	—	14353

Order No Y76

145-8	—	—	14354-7
149	—	1919	14358
150	—	—	14359

Order No Y85

923	—	1917	14438
924	—	1915	14439
925	—	1919	14360
926-7	—	—	14361-2

Order No Y92

136	—	—	14365
137-8	—	—	14363-4

Class 139

139	—	—	14440

Order No Y97

132-5	—	—	14441-4

Class 117
Order No Y101

117-20	—	—	14445-8
121	—	—	Wdn 1915
122	—	—	14449

Class 40
Order No Y105

43-8	—	—	14450-5

Order No Y109

39-42	—	—	14456-9
123	—	—	14460

Class 113
Order No Y113

113-6	—	—	14461-4
121	—	—	14465
124	—	—	14466

Class 928
Order No Y118

928-37	—	—	14467-76

APPENDIX 1

Eng No	Reno'd	Reb'lt	LMS Nos
	Class 72		
	Order No Y124		
72-81	—	—	14477-86
82-91	—	—	14487-96
66-71	—	—	14497-502
92-7	—	—	14503-8
	Class 294		
	Neilson		
294	1294	1915	17249
295	262	1918	17232
296	263	1904	17233
297	264	—	17234
298	539	—	17246
299-300	259-60	—	17230-1
301	335	—	17235
302	337	—	17236
303	365	1902	17241
304	367	1916	17242
305	374	—	17243
306	403	1917	17244
307	517	—	17245
308	548	1917	17247
	RA		
349	—	1902	17237
350	—	—	17238
351	—	1916	17239
352	—	1918	17240
353-4	—	—	17250-1
	Neilson		
517	1517	—	17269
518	—	1917	17252
519-20	—	—	17253-4
521	—	1916	17255
522	—	1917	17256
523	—	1905	17257
524	—	1906	17258
525	—	1921	17259
526	—	1917	17260
680	1680	1899	17270
681	—	1916	17261
682	1682	1910	17271
683	—	1916	17262
684	—	—	17263
685	—	1903 1916	17264
686	—	1917	17265

Eng No	Reno'd	Reb'lt	LMS Nos
687	—	1922	17266
688	—	1907 1920	17267
689	—	1918	17268
	SW		
690	—	1905	17276
691-2	361-2	—	17272-3
693-4	363-4	1916	17274-5
695	365 1365	—	17294
	Order No Y5		
309	549	—	17248
310	1310	—	17290
311	553	—	17283
312	558	1907	17284
313	680	1916	17285
314	682	1915	17286
315	1315	1913	17291
316	703	1906	17287
317	705	—	17288
318	1318	1910	17292
319	1319	—	17293
320	707	—	17289
	Order No Y9		
355	—	—	17277
356	—	1910	17278
357	—	1908	17279
358	—	1916	17280
359	—	—	17281
360	—	1908	17282
366	—	—	17304
367	1367	1895	17312
368-9	—	—	17305-6
370	—	1920	17307
371	—	—	17308
	Order No Y12		
321	708	1905	17310
322	380	1917	17309
339	323 1323	—	17311
340	—	—	17295
341	—	1903	17296
342	—	1899	17297
343-4	—	1916	17298-9
345	—	1913	17300
346	—	1907	17301
347	—	1905 1917	17302
348	—	1906	17303

APPENDIX 1

Eng No	Reno'd	Reb'lt	LMS Nos	Eng No	Reno'd	Reb'lt	LMS Nos
Order No Y23				*Order No Y37*			
403	1403	—	17313	256	—	—	17377
404	—	—	17314	257	—	1918	17378
406	—	1917	17315	258	—	—	17379
407	—	—	17316	259	1259	—	17389
408	—	1916	17317	260	1260	—	17390
409	—	—	17318	261	—	—	17380
410	—	1915	17319	334	—	—	17381
411	—	1918	17320	335	1335	—	17391
412	—	—	17321	336	—	—	17382
413	—	1912	17322	337	1337	—	17392
414	—	1910	17323	203	338	1916	17383
415	—	—	17324	204	339	1921	17384
Order No Y29				*Order No Y38*			
372	—	1918	17325	703	1703 1704	—	17353
373	—	—	17326	704	—	1909	17385
374	1374	1909	17348	705	1705 1706	—	17354
375	—	1909	17327	706	—	—	17386
376	—	—	17328	707	1707	—	17355
377	—	1907	17329	708	1708	—	17356
378	—	—	17330	709-10	—	—	17387-8
379	—	1917	17331				
540-3	—	—	17332-5	**Class 711**			
Order No Y30				*Order No Y41 sup*			
691-4	—	—	17342-5	711-20	—	—	17393-402
695	—	1922	17346	*Order No Y45*			
696	—	—	17347	736-7	—	—	17403-4
Order No Y31				738	—	1917	17405
544	—	—	17336	739-45	—	—	17406-12
545	—	1921	17337	746	—	1921	17413
546	—	1922	17338	747-60	—	—	17414-27
547	—	1907	17339	*Order No Y46*			
548	1548	—	17349	564	—	1917	17433
549	1549	—	17350	565	—	—	17434
550	—	—	17357	566	—	1918	17435
551-2	—	—	17340-1	567-75	—	—	17436-44
553	1553	—	17351	*Order No Y47*			
Order No Y32				576-7	—	—	17445-6
554-7	—	—	17358-61	578	—	1916	17447
558	1558	—	17352	579	—	1917	17448
559-63	—	—	17362-6	580-2	—	—	17449-51
Order No Y33				583-7	—	—	17469-73
697	—	1917	17367	*Order No Y49*			
698-702	—	—	17368-72	761	—	1920	17464
Order No Y36				762-5	—	—	17465-8
199	—	1918	17373	588-92	—	—	17452-6
200-2	—	—	17374-6				

APPENDIX 1

Eng No	Reno'd	Reb'lt	LMS Nos
Order No Y50			
329-33	—	—	17428-32
593-7	—	—	17457-61
598-9	—	1918	17462-3

Class 812
Order No Y54

812-28	—	—	17550-66

Order No Y58

282-4	—	—	17617-9
285	—	1921	17620
286-9	—	—	17621-4
290	—	1908	17625
291	—	—	17626
292	—	1920	17627
293	—	—	17628

Neilson Reid

829	—	1908	17567
830-2	—	—	17568-70
833	—	1922	17571
834-42	—	—	17572-80
843	—	1921	17581
844-8	—	—	17582-6

Sharp Stewart

849-54	—	—	17587-92
855	—	1921	17593
856-9	—	—	17594-7
860	—	1922	17598
861-3	—	—	17599-601

Dübs

864-78	—	—	17602-16

Class 652
Order No Y86

654	—	—	17631
658	—	1921	17635
659	—	—	17636
662-4	—	—	17637-9

Order No Y87

652-3	—	—	17629-30
656-7	—	—	17633-4
665	—	—	17640

Order No Y89

325-8	—	—	17641-4
423	655	—	17632
460	661	—	17645

Class 30
Order No Y98

30-3	—	—	17646-9

Class 34
Order No Y103

34-8	—	—	17800-4

Class 300
Order No Y119

300-9	—	—	17656-65

Order No Y120

310-21	—	—	17666-77

Order No Y122

322-4	—	—	17678-80
294-9	—	—	17650-5

Order No Y123

280-1	—	—	17681-2
670-9	—	—	17683-92

Class 55
Order No Y66

55-9	—	—	14600-4

Order No Y75

51-4	—	—	14605-8

Class 49
Order No Y69

49-50	—	—	14750-1

Class 903
Order No Y80

903-6	—	—	14752-5
907	—	—	Wdn 1915

Class 918
Order No Y79

918-22	—	—	17900-4

Class 908
Order No Y81

908-17	—	—	14609-18

APPENDIX 1

Eng No	Reno'd	Reb'lt	LMS Nos	Eng No	Reno'd	Reb'lt	LMS Nos
	Class 179			269	—	—	Wdn
				270	—	—	16011
	Order No Y107			271	1271	—	16014
179-83	—	—	17905-9		*Order No Y22*		
	Order No Y112			615-8	—	—	16015-8
184-9	—	—	17910-5	619	—	—	Wdn 1920
	Class 60			620	—	—	16019
					Order No Y27		
	Order No Y115			510-5	1510-5	—	16020-5
60	—	—	14650		*Order No Y43*		
	Order No Y116			611-4	—	—	16026-9
61-5	—	—	14651-5		*Order No Y63*		
	Class 938			621-6	—	—	16030-5
					Order No Y68		
938-43	—	—	17756-61	627-8	—	—	16036-7
	Class 956				*Order No Y88*		
	Order No Y125			431	269	—	16039
956-9	—	—	14800-3	463	265	—	16038
	Class 191				**Class 171**		
					SN		
	Order No Y128			171	1171	—	Wdn 11/21
191-4	—	—	14619-22	172	1172	1904	15100
	Order No Y129			173	1173	—	Wdn 8/22
195-8	—	—	14623-6	174	1174	1904	15101
	Class 600			175	1175	1901	15102
				176	1176	—	Wdn 3/21
	Order No Y65			177-8	1177-8	—	15103-4
600-1	—	—	17990-1	228	1228	—	Wdn 9/21
	Order No Y67			229	1229	—	15105
602-3	—	—	17992-3	230	1230	—	Wdn 10/22
	Order No Y70			231	1231	—	Wdn 9/22
604-7	—	—	17994-7		*Order No Y19*		
	Class 262			222-4	1222-4	—	15106-8
				225	1225	—	Wdn 12/23
	Order No Y1			226	1226	—	Wdn 5/21
262-3	1262-3	—	15000-1	227	1227	—	15109
	Class 264				*Order No Y26*		
				189	1189	—	Wdn 4/17
	Order No Y1 (cont)			190	—	—	15110
264-5	1264-5	—	16012-3	191-4	1191-4	—	15111-4
266-8	—	—	16008-10		**Class 1**		
					Order No Y34		
				1-12	—	—	15020-31

192

APPENDIX 1

Eng No	Reno'd	Reb'lt	LMS Nos
Class 19			
Order No Y40			
19-28	—	—	15115-24
Class 92			
Order No Y48			
92-7	13-8	—	15125-30
98-103	—	—	15131-6
Class 879			
Order No Y59			
879-86	—	—	15137-44
Order No Y60			
437-8	—	—	15145-6
Class 104			
Order No Y56			
104-11	—	—	15147-54
167-70	—	—	15155-8
Class 439			
Order No Y61			
439-43	—	—	15159-63
Order No Y64			
444-55	—	—	15164-75
Order No Y77			
151	—	—	15176
473	—	—	15178
655	423	—	15177
Order No Y78			
112	—	—	15179
125	—	—	15180
424	—	—	15181
384	463	—	15182
464-6	—	—	15183-5
660	—	—	15186
666	—	—	15187
Order No Y84			
158	—	—	15188
419	—	—	15189
422	—	—	15190
429	—	—	15191
470	—	—	15192

Eng No	Reno'd	Reb'lt	LMS Nos
Order No Y90			
126-7	—	—	15193-4
420-1	—	—	15195-6
427-8	—	—	15197-8
456	—	—	15199
467-9	—	—	15200-2
Order No Y94			
155	—	—	15203
160	—	—	15204
459	—	—	15205
461	—	—	15206
Order No Y96			
152-4	—	—	15207-9
156	—	—	15210
380	460	—	15211
462	—	—	15212
Order No Y102			
157	—	—	15213
383	164	—	15214
457-8	—	—	15215-6
Order No Y106			
228-31	—	—	15217-20
Order No Y110			
222-4	—	—	15221-3
Order No Y114			
225-7	—	—	15224-6
159	—	—	15227
161-3	—	—	15228-30
Order No Y130			
418	—	—	15231
425-6	—	—	15232-3
430	—	—	15234
Class 431			
Order No Y130 (cont)			
431-4	—	—	15237-40
Class 439			
Order No Y130 (cont)			
435-6	—	—	15235-6
Class 272			
Order No Y16			
272	1272	—	Wdn 3/22
273-4	1273-4	—	16100-1

C.R.—M

APPENDIX 1

Eng No	Reno'd	Reb'lt	LMS Nos
508	1508	—	Wdn 8/22
509	1509	—	16102
527	1527	—	Wdn 12/21

Class 385

Order No Y14

323	505	—	16205
385	—	—	16204
506-7	—	—	16206-7
538	384	—	16203
539	383	—	16202

Class 232

Order No Y18

232-3	1232-3	—	16212-3
234	1234	—	Wdn 12/22
235	1235	—	16214
216-7	—	—	16200-1

Order No Y20

218-20	—	—	16208-10
221	—	1902	16211
386	—	—	Wdn
387	—	—	Wdn

Order No Y24

388-9	—	—	16215-6
391-2	—	—	16217-8
394	1394	—	Wdn 9/22
395	1395	—	16224
396	1396	—	Wdn
398-402	—	—	16219-23

Class 211

Order No Y39

211-5	—	—	16225-9

Class 29

Order No Y42

29	—	—	16231
203-10	—	—	16232-9

Class 782

Order No Y52

782-8	—	—	16254-60

Order No Y53

789-811	—	—	16261-83

Order No Y55

Eng No	Reno'd	Reb'lt	LMS Nos
236-47	—	—	16240-51
485	—	—	16252
516	—	—	16253

Order No Y73

631-40	—	—	16285-94

Order No Y74

641-9	—	—	16295-303
650	—	—	16230

Order No Y82

128-9	—	—	16304-5
166	—	—	16306
324	271	—	16307
425	275	—	16308
434	489	—	16312
472	—	—	16311
501	—	—	16313
630	—	—	16315
668	—	—	16317

Order No Y83

417	—	—	16310
426	276	—	16309
629	—	—	16314
667	—	—	16316
669	—	—	16318

Order No Y91

432	487	—	16325
433	488	—	16284
475	—	—	16319

Order No Y93

435-6	490-1	—	16326-7
476-80	—	—	16320-4

Order No Y95

416	—	—	16328
418	482	—	16331
474	—	—	16329
481	—	—	16330
483-4	—	—	16332-3
500	—	—	16334
608-10	—	—	16335-7

Order No Y99

277	—	—	16338
390	—	—	16344

APPENDIX 1

Eng No	Reno'd	Reb'lt	LMS Nos	Eng No	Reno'd	Reb'lt	LMS Nos
393	—	—	16339		**Class 498**		
397	—	—	16340		*Order No Y100*		
405	—	—	16341				
430	486	—	16343	498-9	—	—	16151-2
471	—	—	16342		*Order No Y111*		
	Order No Y104			527-32	—	—	16153-8
130-1	—	—	16345-6		*Order No Y121*		
171-4	—	—	16347-50	533-8	—	—	16159-64
651	—	—	16351		*Order No Y126*		
781	—	—	16352	502-4	—	—	16165-7
	Order No Y108			510-5	—	—	16168-73
175-8	—	—	16353-6				
508-9	—	—	16357-8		**Class 944**		
	Order No Y117				*NBL*		
232-5	—	—	16359-62	944-55	—	—	15350-61
272-4	—	—	16363-5		**Class 492**		
	Order No Y127				*Order No Y71*		
248-55	—	—	16366-73				
394-6	—	—	16374-6	492-7	—	—	16950-5

APPENDIX 2

BOILER DIMENSIONS

Order Nos	Max ext dia	Length: Between tubeplates	Length: Outer FB casing	Depth of FB below BCL: Front	Depth of FB below BCL: Rear	Working pressure lb/sq in	Tubes	Heating surface sq ft: THS	Heating surface sq ft: FBHS	Heating surface sq ft: SHS	Grate area sq ft
Y1/16/22/27/43/63/68/88	3ft 9¾in	10ft 0in	3ft 6in	3ft 6in	3ft 6in	140	138×1¾in	632	52	–	10.23
SN, Y19/26	3ft 10in	9ft 0$\frac{11}{16}$in	4ft 8¼in	4ft 6¼in	4ft 6¼in	150	138×1¾in	572.5	65	–	14
Y100/111/121/126	4ft 4⅛in	8ft 10in	4ft 9in	4ft 6in	4ft 0in	160	210×1¾in	786	74	–	14.25
Y13	4ft 4⅛in	10ft 4in	5ft 5in	5ft 6in	5ft 0in	150	177×1¾in	837.5	99	–	16.75
Y14/18/20/24	4ft 4⅛in	10ft 4in	5ft 5in	5ft 6in	5ft 0in	150	177×1¾in	837.5	101.1	–	17
Y28	4ft 4⅛in	10ft 4in	5ft 5in	5ft 6in	5ft 0in	160	234×1⅝in	1028.5	111.2	–	17
Y34/39/40/48	4ft 4⅛in	10ft 4in	5ft 5in	5ft 6in	5ft 0in	150	224×1⅝in	984.6	111.2	–	17
Y42/52-3/55-6/59/60-1/64	4ft 4⅛in	10ft 4in	5ft 5in	5ft 6in	5ft 0in	150	206×1¾in	975	110.9	–	17
Y73-4/77-8/82-4/90-1/93-6/99/102/104/106/108/110/114/117/127	4ft 4⅛in	10ft 4in	5ft 5in	5ft 6in	5ft 0in	160	210×1¾in	993.6	110.9	–	17
Y130	4ft 4⅛in	10ft 4in	5ft 5in	5ft 6in	5ft 0in	180	210×1¾in	993.6	110.9	–	17
Engine No 123	4ft 4⅛in	10ft 7in	5ft 6in	5ft 6in	5ft 0½in	150	196×1¾in	950.3	103	–	17.25
E561-2/580, RA SW Y						150	216×1¾in	1095.6	113	–	19.5

Y25	4ft 6¼in	10ft 7in	6ft 2⅜in	5ft 3in	4ft 9in	160	216 × 1¾in 10 × 1½in	109.56	113	—	19.5
Y29/30-1	4ft 6¼in	10ft 7in	6ft 2⅜in	5ft 3in	4ft 9in	150	242 × 1⅝in	1089.7	112.6	—	19.5
Y32-3/36-8	4ft 6¼in	10ft 7in	6ft 2⅜in	5ft 3in	4ft 9in	150	238 × 1⅝in	1071.5	112.6	—	19.5
Y35	4ft 6¼in	10ft 7in	6ft 2⅜in	5ft 3in	4ft 9in	160	238 × 1⅝in	1071.5	112.6	—	19.5
Y41/45-7/49/50	4ft 6¼in	10ft 7in	6ft 2⅜in	5ft 3in	4ft 9in	150	218 × 1¾in	1056.8	112.4	—	19.5
Y71	4ft 6¼in	10ft 7in	6ft 2⅜in	5ft 3in	4ft 9in	175	222 × 1¾in	1076.3	112.7	—	19.5
Y44/54/58/86-7/89	4ft 9¼in	10ft 7in	6ft 5in	5ft 6in	5ft 0in	160	265 × 1¾in	1284.5	118.8	—	20.63
Y98/103	4ft 9¼in	10ft 7in	6ft 5in	5ft 6in	5ft 0in	160	161 × 1¾in 21 × 5in	1071.3	118.8	266.9	20.63
Y119/120/122-3	4ft 9¼in	10ft 7in	6ft 5in	5ft 6in	5ft 0in	170	275 × 1¾in	1332.9	118.8	—	20
Y51	4ft 9¼in	11ft 4½in	6ft 5in	5ft 6in	5ft 0in	175	265 × 1¾in	1381.2	118.8	—	20.63
Y51	4ft 9¼in	11ft 4½in	6ft 5in	5ft 6in	5ft 0in	170	159 × 1¾in 18 × 5in	1094.3	118.8	214	20.63
Y57/62	4ft 9¼in	11ft 4½in	6ft 11in	5ft 9in	5ft 3in	180	269 × 1¾in	1402	138	—	23
Y57/62	4ft 9¼in	11ft 4½in	6ft 11in	5ft 9in	5ft 3in	170	159 × 1¾in 18 × 5in	1094.3	138	214	23
Y65/67/70	4ft 9¼in	15ft 7½in	6ft 11in	5ft 9in	5ft 3in	175	275 × 1¾in	1970	138	—	23
Y66	4ft 9¼in	14ft 3½in	6ft 5in	5ft 0in	4ft 3in	175	275 × 1¾in	1800	105	—	20.63
Y75	4ft 9¼in	13ft 6⅝in	6ft 5in	5ft 0in	4ft 3in	175	275 × 1¾in	1707	105	—	20.63
Y128-9	4ft 9¼in	13ft 6⅝in	7ft 0in	5ft 0in	4ft 3in	185	275 × 1¾in	1707	116	—	21.9
NBL 4-6-2T	4ft 9¼in	14ft 6in	7ft 0in	5ft 0in	4ft 2in	170	159 × 1¾in 18 × 5in	1395	121	200	21.5
Y72/76/85/92	5ft 0in	11ft 6in	7ft 0in	6ft 0in	5ft 6in	180	255 × 1¾in 21 × 2in	1470	145	—	21
Y97/101	5ft 0in	11ft 6in	7ft 0in	6ft 0in	5ft 6in	165	163 × 1¾in 24 × 5in	1220	145	330	21

Order Nos	Max ext dia	Length Between tubeplates	Length Outer FB casing	Depth of FB below BCL Front	Depth of FB below BCL Rear	Working pressure lb/sq in	Tubes	Heating surface sq ft THS	Heating surface sq ft FBHS	Heating surface sq ft SHS	Grate area sq ft
Y105/9	5ft 0in	11ft 6in	7ft 0in	6ft 0in	5ft 6in	170	163 × 1¾in 24 × 5in	1220	145	295	21
Y113/118	5ft 0in	11ft 6in	7ft 0in	6ft 0in	5ft 3in	175	157 × 1¾in 24 × 5in	1185	144	200	20.7
Y124	5ft 0in	11ft 6in	7ft 0in	6ft 0in	5ft 3in	180	157 × 1¾in 24 × 5in	1185	144	200	20.7
Y69	5ft 0in	17ft 3in	8ft 6in	5ft 3in	3ft 9in	200	257 × 1¾in 13 × 2½in	2178	145	–	26
Y69	5ft 0in	15ft 8in	8ft 6in	5ft 3in	3ft 9in	200	230 × 2in	1887	145	–	26
Y69	5ft 0in	15ft 8in	8ft 6in	5ft 3in	3ft 9in	175	125 × 2in 24 × 5in	1509	145	515	26
938 class	5ft 3in	14ft 9⅛in	8ft 0in	5ft 6in	4ft 4in	160	129 × 2in 24 × 5in	1460	139.6	350	25.3
Y79	5ft 3½in	13ft 6½in	6ft 11in	5ft 0in	4ft 3in	175	242 × 2in	1895	128	–	21
Y81	5ft 3½in	14ft 11 7/16in	6ft 11in	5ft 0in	4ft 3in	180	242 × 2in	2050	128	–	21
Y107/112	5ft 3½in	14ft 11 7/16in	6ft 11in	5ft 0in	4ft 3in	170	132 × 2in 24 × 5in	1439	128	403	21
Y80	5ft 3½in	16ft 8in	8ft 6in	5ft 0in	3ft 9in	200	242 × 2in	2117.5	148.3	–	26
Y80	5ft 3½in	16ft 8in	8ft 6in	5ft 0in	3ft 9in	175	132 × 2in 24 × 5in	1666	148.3	515	26
Y115-6	5ft 3½in	15ft 3in	8ft 6in	5ft 0in	3ft 11in	175	130 × 2in 24 × 5in	1529.5	146.5	258.3	25.5
Y125	5ft 9in	16ft 0in	9ft 3in	5ft 3in	4ft 3in	180	203 × 2in 24 × 5in	2200	170	270	28

** See text

APPENDIX 3

PORT DIMENSIONS AND VALVE EVENTS: SLIDE-VALVE ENGINES

Class	Cylinders	Length of ports	Breadth of ports		Lap of valve	Lead	Travel in full gear
			Steam	Exhaust			
264	14in x 20in	11in	$1\frac{1}{4}$in	$2\frac{1}{2}$in	$\frac{3}{4}$in		
171	16in x 22in	13in	$1\frac{1}{4}$in	2in			
498	17in x 22in	13in	$1\frac{1}{2}$in	3in	1in	$\frac{3}{16}$in front $\frac{3}{32}$in rear	$3\frac{29}{32}$in
1, 104	17in x 24in	15in	$1\frac{3}{8}$in	3in			
66 (batch Y)	18in x 26in	8in+8in	$1\frac{5}{8}$in	$3\frac{1}{4}$in	1in	$\frac{3}{16}$in	$4\frac{3}{16}$in
80, 123		8in+8in	$1\frac{5}{8}$in	$3\frac{1}{4}$in	1in		
294, 19		8in+8in	$1\frac{5}{8}$in	$3\frac{1}{4}$in	1in		
29, 92, 439		8in+8in	$1\frac{5}{8}$in	$3\frac{1}{4}$in	1in		
782, 879		8in+8in	$1\frac{5}{8}$in	$3\frac{1}{4}$in	1in		
76 (Y21)	18in x 26in	8in+8in	$1\frac{5}{8}$in	$3\frac{1}{4}$in+$3\frac{1}{4}$in	1in	$\frac{1}{8}$in	$4\frac{3}{16}$in
13	18in x 26in	8in+8in	$1\frac{5}{8}$in	$3\frac{1}{4}$in	1in	$\frac{1}{4}$in front $\frac{1}{32}$in rear	$4\frac{5}{16}$in
721	$18\frac{1}{4}$in x 26in	8in+8in	$1\frac{5}{8}$in	$3\frac{1}{4}$in	1in	$\frac{3}{16}$in	$4\frac{3}{16}$in
812	$18\frac{1}{2}$in x 26in	18in	$1\frac{5}{8}$in	$3\frac{1}{4}$in	$1\frac{1}{16}$in		
280 (Y123)	$18\frac{1}{2}$in x 26in	15in	$1\frac{3}{4}$in	$3\frac{1}{2}$in	1in	0in front $\frac{3}{16}$in rear	$4\frac{7}{8}$in
124 (Dübs)	19in x 26in	16in	$1\frac{5}{8}$in	$3\frac{1}{4}$in	1in	$\frac{1}{8}$in	$4\frac{1}{2}$in
766, 900	19in x 26in	18in	$1\frac{5}{8}$in	$3\frac{1}{4}$in			
140, 55	19in x 26in	18in	$1\frac{5}{8}$in	$3\frac{1}{4}$in			
908, 918	19in x 26in	18in	$1\frac{5}{8}$in	$3\frac{1}{4}$in			
492	19in x 26in	18in	$1\frac{5}{8}$in	$3\frac{1}{4}$in			
191	$19\frac{1}{2}$in x 26in	$16\frac{1}{2}$in	$1\frac{3}{4}$in	$3\frac{1}{2}$in	$1\frac{1}{4}$in	$\frac{1}{4}$in	$5\frac{1}{2}$in
903	20in x 26in	18in	$1\frac{3}{4}$in	$3\frac{1}{2}$in			
49	21in x 26in	18in	$1\frac{3}{4}$in	$3\frac{1}{2}$in		0in front $\frac{3}{16}$in rear	$4\frac{23}{32}$in
600	21in x 26in	$15\frac{1}{2}$in	2in	4in	1in	0in front $\frac{3}{16}$in rear	$5\frac{3}{16}$in

Steam pipe diameters were as follows:—

	Boiler	Smokebox
1 class	$4\frac{1}{4}$in	$3\frac{1}{4}$in
18in x 26in classes prior to Y21	$4\frac{1}{4}$in	$3\frac{1}{4}$in
104 class	5in	4in
18in x 26in classes from Y21	5in	4in
All larger McIntosh classes	5in	4in

APPENDIX 4

FRAMES AND MOTION

Class	Frames Length	Frames Thickness	Wheelbase Bogie	Wheelbase Coupled	Wheelbase Total	Cylinders Centre to centre	Lengths Centre to centre Connecting rods	Lengths Centre to centre Eccentrics	Wheel diameter Bogie	Wheel diameter Driving	Driving journals Diameter	Driving journals Length
294	24ft 9in	1in	–	7ft 6in + 8ft 9in	16ft 3in	2ft 3in	6ft 6in	5ft 6in	–	5ft 0in	8in	7½in
711	24ft 9in	1in	–	7ft 6in + 8ft 9in	16ft 3in	2ft 3in	6ft 6in	4ft 7in	–	5ft 0in	8in	7½in
812	25ft 3in	1in	–	7ft 9in + 9ft 0in	16ft 9in	2ft 4½in	6ft 6in	4ft 11in	–	5ft 0in	8½in	7½in
30	27ft 6in	1in	–	7ft 9in + 9ft 0in	16ft 9in	2ft 1½in	6ft 6in	4ft 2in	–	5ft 0in	8½in	9in
34	30ft 0in	1in	–	7ft 9in + 9ft 0in	24ft 0in	2ft 1½in	6ft 6in	4ft 2in	–	5ft 0in	8½in	9in
300	25ft 9in	1⅛in	–	7ft 9in + 9ft 0in	16ft 9in	2ft 1½in	6ft 6in	4ft 2in	–	5ft 0in	8½in	9in
191	32ft 4in	1⅛in	3ft 5in + 3ft 7in	6ft 2in + 6ft 2in	24ft 9in	6ft 9¾in	10ft 3in	4ft 7⅛in	3ft 6in	5ft 6in	8in	9⅛in
55	31ft 0in	1in	5ft 9in	5ft 3in + 6ft 0in	23ft 9½in	2ft 4½in	6ft 6in	4ft 7in	3ft 6in	5ft 0in	8½in	7½in
918	31ft 6in	$1\frac{1}{16}$in	5ft 9in	5ft 3in + 6ft 0in	23ft 9½in	2ft 4½in	6ft 6in	4ft 7in	3ft 6in	5ft 0in	9¼in	7½in
908	32ft 11in	$1\frac{1}{16}$in	5ft 9in	6ft 8in + 6ft 8in	25ft 10½in	2ft 4½in	6ft 6in	4ft 7in	3ft 6in	5ft 9in	9¼in	7½in
179	33ft 6in	1⅛in	5ft 9in	6ft 8in + 6ft 8in	26ft 1½in	2ft 1½in	6ft 6in	4ft 2in	3ft 6in	5ft 9in	9½in	9in
49	37ft 0in	1⅛in	3ft 2in + 3ft 4in	7ft 6in + 7ft 6in	28ft 8in	2ft 1½in	6ft 8in	4ft 6in	3ft 6in	6ft 6in	9½in	9½in
903	37ft 0in	$1\frac{1}{16}$in	3ft 2in + 3ft 4in	7ft 2in + 7ft 6in	28ft 8in	2ft 0¼in	7ft 0in	4ft 6in	3ft 6in	6ft 6in	9½in	10½in
60	36ft 0in	1¼in	3ft 5in + 3ft 7in	7ft 0in + 7ft 6in	27ft 6in	6ft 9¾in	11ft 0in	5ft 3in	3ft 6in	6ft 1in	9½in	11in
956	37ft 8in	1¼in	3ft 5in + 3ft 7in	7ft 0in + 8ft 0in	28ft 8in	6ft 9¾in	11ft 0in 6ft 6in	 3ft 11in*	3ft 6in	6ft 1in	9½in	11in
938	33ft 7½in	1⅛in	6ft 6in	6ft 3in + 8ft 0in	26ft 4½in	6ft 8½in	10ft 3½in	4ft 7⅝in	3ft 3in	6ft 0in	8½in	10in

600	31ft 8in	1⅛in		8ft 6in+5ft 4in+8ft 6in	22ft 4in	2ft 1½in	7ft 1in	4ft 7in	–	4ft 6in	8½in	8in
123	26ft 8in	1in	6ft 6in	–	21ft 1in	2ft 3in	6ft 6in	4ft 7in	3ft 6in	7ft 0in	8½in	7½in
80	26ft 6in	1in	6ft 6in	8ft 0in	21ft 1in	2ft 3in	6ft 6in	5ft 6in	3ft 6in	5ft 9in	8in	7½in
66	27ft 11in	1in	6ft 6in	9ft 0in	22ft 1in	2ft 3in	6ft 6in	4ft 7in	3ft 6in	6ft 6in	8in	7½in
13	27ft 11in	1in	3ft 2in+3ft 4in	9ft 0in	22ft 1in	2ft 3in	6ft 6in	5ft 6in	3ft 6in	6ft 6in	8in	7½in
721	27ft 11in	1in	3ft 2in+3ft 4in	9ft 0in	22ft 1in	2ft 3in	6ft 6in	4ft 7in	3ft 6in	6ft 6in	8in	7½in
766	28ft 11in	1in	3ft 2in+3ft 4in	9ft 0in	23ft 1in	2ft 4½in	7ft 1in	5ft 6in	3ft 6in	6ft 6in	8½in	7½in
900	29ft 5in	1in	3ft 2in+3ft 4in	9ft 6in	23ft 7in	2ft 4½in	7ft 1in	5ft 6in	3ft 6in	6ft 6in	8½in	7½in
140	29ft 8in	1in	3ft 2in+3ft 4in	9ft 9in	23ft 10in	2ft 4½in	7ft 1in	5ft 6in	3ft 6in	6ft 6in	9¼in	8in
139	29ft 8in	$1\frac{1}{16}$in	3ft 2in+3ft 4in	9ft 9in	23ft 10in	2ft 4½in	7ft 1in	4ft 6in	3ft 6in	6ft 6in	9¼in	7½in
113	30ft 2in	1¼in	3ft 5in+3ft 7in	9ft 9in	24ft 4in	2ft 1½in	7ft 1in	4ft 6in	3ft 6in	6ft 6in	9½in	9in
264	19ft 8in	⅞in	–	7ft 0in	7ft 0in	6ft 4½in	7ft 4½in		–	3ft 8in	6¼in	7½in
171	28ft 9in	⅞in	5ft 0in	6ft 4in	21ft 4in	2ft 3in		4ft 6in	2ft 6in	5ft 0in	6¾in	7½in
1	33ft 0in	1in	6ft 0in	9ft 0in	24ft 10in	2ft 3in	6ft 6in	5ft 6in	3ft 2in	5ft 0in	8in	7½in
104	28ft 6in	1in	5ft 0in	7ft 6in	20ft 6in	2ft 4½in	6ft 6in	4ft 7in	2ft 6in	4ft 6in	8in	7½in
19, 92 & 439	30ft 10in	1in	5ft 6in	7ft 6in	22ft 0in	2ft 3in	6ft 6in	4ft 7in	3ft 2in	5ft 9in	8in	7½in
232	28ft 9in	1in	–	7ft 6in+8ft 9in	16ft 3in	2ft 3in	6ft 6in		–	4ft 6in		
211	28ft 3in	1in	–	7ft 6in+8ft 9in	16ft 3in	2ft 3in	6ft 6in		–	4ft 6in		
29 & 782	27ft 9in	1in	–	7ft 6in+8ft 9in	16ft 3in	2ft 3in	6ft 6in	4ft 7in	–	4ft 6in	8in	7½in
492	31ft 0in	1in	–	7ft 9in+5ft 7½in+5ft 7½in	19ft 0in	2ft 4½in	6ft 6in	4ft 11in	–	4ft 6in	8½in	7½in
498	24ft 2in	1in	–	5ft+5ft	10ft 0in	6ft 7in	5ft 7in	3ft 9in	–	4ft 0in	7in	8in
944	40ft 0in	1¼in	3ft 5in+3ft 7in	6ft 7½in+6ft 7½in	33ft 1in	6ft 9¾in	10ft 6in	4ft 11in	3ft 6in	5ft 9in	9½in	11in

* Inside (Nos 957-8)

APPENDIX 5

MAXIMUM DIMENSIONS

Class	Height above rail: chimney	Height above rail: boiler centre line	Width over running-plate	Width over cab sides	Length over buffers	Total wheelbase (engine & tender)	Weight in WO: Adhesive T cwt	Weight in WO: Total T cwt	Tractive effort (at 85% WP) lb
123	12ft 11in	7ft 6in	7ft 3in	6ft 0⅛in	51ft 9¼in	42ft 5½in	16 9¼	41 7¼	12,790
80	12ft 11in	7ft 3in	7ft 3in	6ft 0⅛in	51ft 7in	42ft 3½in	28 3¼	42 7¼	15,566
66	12ft 11in	7ft 3in	7ft 3in	6ft 0⅛in	53ft 0¼in	43ft 8½in	30 8	45 3	13,770
					54ft 6in	44ft 3½in			
13	12ft 11in	7ft 3in	7ft 3in	6ft 0⅛in	53ft 9¾in	44ft 5½in	30 11¾	45 5¼	14,690
721	12ft 11½in	7ft 9in	7ft 3in	6ft 0⅛in	53ft 9¾in	44ft 5½in	31 5	46 19¾	15,100
766	12ft 11½in	7ft 9in	7ft 8in	6ft 8½in	57ft 3¾in	49ft 2½in	32 14	49 0	17,900
766 (sup)	12ft 11½in	8ft	7ft 8in	6ft 8½in	57ft 3¾in	49ft 2½in	34 15	52 0	18,315
900	12ft 11½in	8ft	7ft 8in	6ft 8½in	57ft 3¾in	49ft 2½in	34 13	51 14	18,411
900 (sup)	12ft 11½in	8ft	7ft 8in	6ft 8½in	57ft 3¾in	49ft 2½in	36 5	54 10	18,315
140	12ft 11in	8ft 3in	8ft	7ft	58ft 2in	49ft 5½in	37 15	56 10	18,411
139	12ft 11in	8ft 3in	8ft	7ft	58ft 2in	49ft 5½in	38 0	59 0	18,700
40	12ft 11in	8ft 3in	8ft	7ft	58ft 2in	49ft 5½in	38 0	59 0	19,751
113	12ft 10½in	8ft 3in	8ft	7ft	56ft 2in	46ft 8½in	39 15	61 5	19,833
72	12ft 10½in	8ft 3in	8ft	7ft	56ft 2in	46ft 8½in	39 15	61 5	21,435
294	12ft 11in	7ft 3in	7ft 3in	6ft 0⅛in	49ft 10¾in	37ft 4½in	41 6	41 6	17,901
711	12ft 11in	7ft 3in	7ft 3in	6ft 0⅛in	49ft 10¾in	37ft 4½in	40 6*	40 6*	17,901
812	12ft 11in	7ft 9in	7ft 8in	6ft 6in	51ft 1¾in	38ft 7½in	45 13¾	45 13¾	20,169
30	12ft $9\frac{3}{16}$in	8ft 3in	7ft 8in	6ft 6in	53ft 6in	38ft 7½in	51 2½	51 2½	22,409
34	12ft $9\frac{3}{16}$in	8ft 3in	7ft 8in	6ft 6in	56ft	45ft 10½in	46 1	54 5	22,409
300	12ft 10in	8ft 3in	7ft 8in	6ft 6in	51ft 9in	38ft 7½in	49 5	49 5	21,430
191	12ft 10in	8ft 6in	8ft 10in	7ft 3in	55ft 5in	46ft 4½in	45 16½	62 15½	23,555
55	12ft 11in	8ft	7ft 8in	6ft 6in	54ft 1¼in	45ft	42 17	57 8	23,269
918	12ft 11in	8ft 6in	7ft 8in	6ft 10in	57ft 6in	48ft	45 18	60 8	23,269

908	12ft 11in	8ft 6in	7ft 8in	6ft 10in	58ft 11in	49ft 5in	49	0	64	0	20,812
179	12ft 11in	8ft 6in	7ft 8¼in	6ft 10¼in	59ft 6in	49ft 8in	51	5	68	10	20,704
49	12ft 11in	8ft 6in	8ft	7ft	65ft 6in	56ft 10½in	55	0	73	0	24,990
49 (1904)	12ft 11in	8ft 6in	8ft	7ft	65ft 6in	56ft 10½in	53	10	70	0	22,667
49 (sup)	12ft 11in	8ft 6in	8ft	7ft	65ft 6in	56ft 10½in	54	10	71	10	21,348
903	12ft 11in	8ft 6in	8ft	7ft 2in	65ft 6in	56ft 10½in	55	0	73	0	22,667
903 (sup)	12ft 11in	8ft 6in	8ft	7ft 2in	65ft 6in	56ft 10½in	55	15	74	5	21,348
60	12ft 10½in	8ft 6in	8ft	7ft	62ft	52ft 5½in	56	10	75	0	21,155
956	12ft 10½in	8ft 10in	8ft 10in	7ft 9in	63ft 8in	54ft 0½in	60	0	81	0	27,980
938	**	8ft 8½in	8ft 9½in	8ft 1in	59ft 6in	49ft 4½in	52	10	72	6	23,320
600	12ft 11½in	8ft	7ft 8in	6ft 6in	57ft 8in	44ft 6½in	60	12¼	60	12¼	31,584
264		**5ft 4½in**	**7ft 9in**	**7ft 2in**	**22ft 3¾in**	**7ft**	**27**	**7½**	**27**	**7½**	10,601
171	12ft 2½in	6ft 6in	8ft	6ft 0¼in	29ft 3¾in	18ft 10in	25	1	37	15½	11,960
1	12ft 11in	7ft 3in	8ft 6in	6ft 6in	33ft 1¼in	21ft 10in	34	17¾	50	6¾	14,730
19	12ft 11in	7ft 3in	8ft 6in	6ft 6in	33ft 11¼in	22ft	32	10¾	53	16	15,566
92	12ft 11in	7ft 3in	8ft 6in	6ft 6in	33ft 11¼in	22ft	33	11¾	55	15	15,566
879	12ft 11in	7ft 3in	8ft 6in	6ft 6in	33ft 11¼in	22ft	32	10¾	53	10¾	15,566
439	12ft 11in	7ft 3in	8ft 6in	6ft 6in	33ft 11¼in	22ft	33	10	53	19	16,603
431	12ft 11in	7ft 3in	8ft 6in	6ft 6in	34ft 4½in	22ft 6in	34	18½	57	17½	19,200
104	12ft 10 3/16in	7ft 3in	8ft 6in	7ft 9in	31ft 7¼in	20ft 6in	32	0	51	2½	16,376
232		6ft 9in	7ft 3in		31ft 10¼in	16ft 3in	43	16¾	43	16¾	19,890
211		6ft 9in	7ft 3in		31ft 1¼in	16ft 3in	46	5¼	46	5¼	19,890
29	12ft 11in	7ft 3in	8ft 6in	6ft 8in	30ft 10¼in	16ft 3in	49	14½	49	14½	19,890
782	12ft 11in	7ft 3in	8ft 6in	6ft 8in	30ft 10¼in	16ft 3in	47	15¼	47	15¼	19,890
498	12ft 8 3/16in	7ft 1in	8ft 6in	6ft 8in	26ft 9¾in	10ft	47	15	47	15	18,014
492	12ft 1in	8ft 4½in	8ft 9in	6ft 9in	34ft 2½in	19ft	62	15¾	62	15¾	25,855
944	12ft 10in	8ft 3in	8ft 9in	7ft 0in	43ft 2½in	33ft 1in	55	1	91	13	20,704

* 42T 4cwt with condensing apparatus.

** 13ft 3¾in as delivered to the Highland Railway.

APPENDIX 6

TENDERS

Type (Cpcty)	Year	Wheelbase	Outside frames Length	Outside frames Depth	Tank Length*	Tank Breadth*	Tank Depth**	Well Length**	Platform Width	Wheels Diameter	Coal T	Weight WO T	Weight WO cwt
2500	1883	13ft	21ft 4in	2ft	18ft	7ft 2⅝in	3ft 5⅛in	5ft 6in	7ft 3in	4ft	4½	31	19
2800	1884	13ft	21ft 4in	2ft	18ft	7ft 2⅝in	3ft 5⅛in	14ft 5in	7ft 3in	4ft	4½	33	18½
2850	1886 (No 123)	13ft	21ft 4in	2ft 7in	18ft	7ft 2½in	3ft 5⅛in	14ft 0½in	7ft 3in	4ft	4½	33	19
2500	1889	13ft	21ft 4in	2ft 4¾in	18ft	7ft 2⅝in			7ft 3in	4ft	4½	33	19¾
2840	1894	13ft	21ft 4in	2ft 4¾in	18ft	7ft 2⅝in	3ft 11¼in	13ft 8in	7ft 3in	4ft	4½	34	17
3130	1887	13ft	21ft 4in	2ft	19ft 3in	7ft 1¼in	3ft 9in	14ft 5in	7ft 3in	4ft	4½	35	8¼
3550	1889	13ft	22ft 10in	2ft 4¾in	20ft 9in	7ft 1¼in	4ft	15ft 10in	7ft 3in	4ft	4½	40	0¼
3570	1894	13ft	22ft 1in	2ft 4¾in	20ft	7ft 1¼in	4ft 4½in	15ft 0¼in	7ft 3in	4ft	4½	39	1½
3570	1901	13ft	22ft 1in	2ft 4¾in	20ft	7ft 1¼in	4ft 4½in	15ft 0¼in	7ft 8in	4ft	4½	41	0½
4200	1916	13ft	22ft 1in	2ft 4¾in	20ft	7ft 10¼in	4ft 9in	14ft 9¼in	8ft	4ft	6	46	10
4500	1921	13ft	22ft 1in	2ft 4¾in	20ft	8ft 7½in		14ft 9¼in	8ft 10in	4ft	5½	48	0
3000	1899	13ft	22ft 1in	2ft 4¾in	20ft	7ft 1¼in	3ft 7½in	15ft 0¼in	7ft 8in	4ft	4½	37	18
3000	1902	11ft	19ft 2in	2ft 4¾in	17ft 1in	7ft 1¼in	4ft 4½in	12ft 7¼in	7ft 8in	4ft	4½	37	6
3000	1922	11ft	19ft 2in	2ft 4¾in	16ft 7in	8ft 1in	4ft	11ft 9¼in	8ft 10in	4ft	4½	37	16½
4000	1915 (938-43)	13ft	21ft 4in		19ft 2¾in	7ft 8¼in	5ft 5½in	14ft	8ft 9½in		6½	47	10
4125	1897	5ft 6in+5ft 9in +5ft 6in	24ft 7in	1ft 5in	22ft 6in	7ft 6¼in	4ft 1½in	17ft 3in	7ft 8in	3ft 6in	4½	49	10
4125	1899	5ft 6in+5ft 6in +5ft 6in	24ft 7in	1ft 5in	22ft 6in	7ft 6¼in	4ft 1½in	17ft 3in	7ft 8in	3ft 6in	4½	49	10
4300	1904	5ft 6in+5ft 6in +5ft 6in	24ft 7in	1ft 5in	22ft 6in	7ft 10¼in	4ft 3¼in	15ft 8¼in	8ft	3ft 6in	4½		
5000	1903	5ft 6in+5ft 6in +5ft 6in	24ft 7in	1ft 5in	22ft 6in	7ft 10¼in	4ft 11¾in	15ft 8¼in	8ft	3ft 6in	5	55	0
5000	1906	5ft 6in+5ft 6in +5ft 6in	24ft 7in	1ft 5in	22ft 6in	7ft 10¼in	4ft 11¾in	15ft 8¼in	8ft	3ft 6in	5	57	0
4600	1910	5ft 6in+5ft 6in +5ft 6in	24ft 7in	1ft 5in	22ft 6in	7ft 10¼in	4ft 11¾in	None	8ft	3ft 6in	6	56	0

* outside dimensions ** inside dimensions

APPENDIX 7

PROJECTED DESIGNS

		4-4-2 1905	4-4-2 1906	4-6-2 1913	2-6-0 1917	2-6-0 1923
Type Year Cylinders	2 LP 2 HP	22in×26in 15in×26in	20in×26in	(4) 16in×26in	21in×28in	21in×28in
Coupled wheel diameter		6ft 6in	6ft 6in	6ft 6in	5ft 8in	5ft 6in
Coupled wheelbase		7ft 6in	7ft 6in	13ft 6in	16ft 6in	16ft 2in
Total wheelbase		28ft 8in	28ft 8in	33ft 10in	26ft 0in	25ft 0in
Boiler: max external diameter		5ft 3in	5ft 3in	5ft 8in	5ft 7¼in	5ft 9in
Length between tubeplates		15ft 8in	16ft 8in	22ft 0in	13ft 3in	11ft 2¼in
Length of outer firebox		8ft 6in	8ft 6in	8ft 6in	9ft 0in	9ft 0in
Heating surface: tubes		1968sq ft	2093sq ft	2440sq ft	1670sq ft	1564sq ft
Heating surface: firebox		163sq ft	163sq ft	158sq ft	165sq ft	168sq ft
Superheating surface		—	—	516sq ft	165sq ft	178sq ft
Grate area		26sq ft	26sq ft	37sq ft	27½sq ft	27sq ft
Working pressure, 1lb/sq in		200	200	180	175	180
Tractive effort at 85% WP, lb		23,250	22,667	24,576	25,420	28,624
Weight WO: adhesion		36¾ tons		55 tons	56¼ tons	60 tons
Weight WO: total		72 tons	71 tons	90 tons	70 tons	72 tons
Length over buffers		65ft 6in	65ft 6in	72ft 3in		59ft 3in

APPENDIX 8

KNOWN ALLOCATIONS CIRCA 1921

Shed	4-6-0s	4-4-0s	0-6-0s	0-4-4Ts	0-6-0Ts	Others
Carlisle	62-3, 906, 957-8, 179, 188-9, 913	47-8, 72-3, 86-8, 95-7, 113-6, 121-4, 134, 136-7, 145-6, 148-9, 896, 898	327, 652, 654, 656-9, 662, 829, 832-6, 843, 846-7, 854, 856-7, 859, 872		173, 233-5, 250-1, 483-4	34-8
Stranraer		1195-6	409, 684, 708			
Beattock				151, 228, 431-4, 441, 443-4, 1194		951-3
Lockerbie				226-7, 419		8
Leadhills				424		
Carstairs		68-9, 78-9, 925-7, 1018, 1061, 1073, 1084	280-1, 314, 323-4, 343, 345, 348, 571, 582, 670, 704, 827, 831, 845, 865	155, 159-60	244-5, 275, 501	
Dalry Road	916, 921	140-1, 143, 889, 899, 923-4, 932-3, 1081, 1124, 732	201, 287-8, 290, 306, 321, 330-3, 548, 554, 661, 672-3, 821-2, 838 862, 869-71, 873, 1548	106-9, 167-70	254, 384, 474, 489, 506, 516, 531, 634, 810-1, 1509	613, 621-2, 625
Grangemouth			337, 340, 415, 517, 523, 685, 703, 1374	17	213-6, 392, 499, 505, 636-7, 645-7, 799, 808	623
Greenhill			263, 346, 363		212	1515
Motherwell	922		257, 259, 297, 335-6, 338, 347, 355-6, 358, 361-2, 365-6, 372-3, 375-6, 521-2, 543,	16, 100-1, 460, 884-6	130, 211, 272-3, 277, 388, 391, 396-7, 508-9, 631, 786, 792-8, 800-6,	492-4, 497, 600, 601, 604, 607

Hamilton		552, 558-60, **568**, 598, 675-6, 683, 688, 691, 693, 737-9, 749-53, 1310, 1315, 1337, 1367, 1680	18, 98-9, 152, 438, 445-7, 451, 454-5, 466, 470, 880-3	236-9, 241-3, 253, 276, 475-8, 487-8, 632, 639, 648-50, 782-5	495, 603
Polmadie	49, 61, 903	142, 727, 766-7, 769-70, 772, 774, 900-2, 936-7, 1079 1090	308-10, 312-3, 334, 339, 344, 349-51, 368, 380, 544, 546, 681, 705, 740-1, 746, 1294, 1319, 1335	105, 111-2, 427, 463-4, 468, 660	128-9, 131, 166, 171, 175-7, 208, 210, 217 220-1, 249, 255, 271, 389, 394, 471-2, 480, 486, 514, 527-9, 533, 609-10, 629-30, 633, 640, 642, 667-9, 787-91	605, 944-5, 955
			30-1, 199, 325, 328, 341, 357, 364, 406, 408, 410-2 556-7, 562, 564-5, 570, 574, 576-9, 599, 677, 679, 689, 696-7, 699, 707, 710, 747, 761-2, 812, 815, 817, 826, 1318			
Kilbirnie				158, 161		
Ardrossan		725	550, 575			
Greenock		723, 731	32-3, 678, 813-4, 818, 863, 874		502-4, 530, 537	265, 614, 946-50, 1512
Dawsholm			256, 264, 329, 369, 404, 413, 524, 583-7, 592, 692, 695, 701-2, 759	21, 24, 28, 102-3, 435-6, 461, 879	174, 209, 417, 510, 532, 535, 641, 644	1, 3-4, 10, 954
Yoker					511, 641	268-9, 611, 628
Dumbarton				15, 19-20, 22		9, 1263
Airdrie				23, 25-7	390, 393, 643	11, 267, **615**, 620

Shed	4-6-0s	4-4-0s	0-6-0s	0-4-4Ts	0-6-0Ts	Others
Balornock	50, 60, 180-3, 197-8, 908-12, 918-9, 938-41, 956	66-7, 76-7, 82-3, 147, 768, 775-8, 890-1, 894-5, 934-5	261, 282-3, 285-6, 289, 291-2, 300-1, 354, 378, 520, 525, 541-2, 549, 553, 596, 744-5, 754-5, 816, 819, 841-2, 1365, 1517	104	203-4, 218, 247, 252, 274, 400-2, 485, 498, 635, 1235	266, 1123
Stirling	53, 57-9, 191-3, 920	773, 1070, 1075	294-5, 304-5, 317-8, 353, 367, 371, 374, 403, 518-9, 539, 547, 551, 597, 671, 690, 720, 756-8, 763	156, 469, 1193	608, 638, 1232, 1273	
Alloa		1114				
Callander		779				
Killin				1175, 1177		
Oban	52, 54-6, 194-6		200			
Perth	64-5, 184-7, 904-5, 917, 942-3, 959	39, 45-6, 74-5, 89-93, 119-20, 132-3, 138-9, 144, 150, 771, 888, 893, 928-31	302, 307, 319-20, 322, 589-91, 748, 824, 839, 860-1, 864, 866-7, 878	153-4, 440, 1223, 1229	172, 383, 399, 416, 482, 781	5, 7
Crieff				448, 458		
Dundee			712, 714-5, 823, 825, 851-3, 858	126, 157, 418		496
Blairgowrie				230		
Forfar			315-6, 572, 580-1, 1403	163-4, 190, 422, 425-6, 428, 452, 462, 465, 1172	1274	
Brechin				456, 467		
Aberdeen		40-4, 70-1, 80-1, 84-5, 117-8, 726, 897	298-9, 303, 326, 711, 718, 828, 849-50, 855		232, 491	

APPENDIX 9

MILEAGES

Class		Greatest Engine No	Miles	Least Engine No	Miles
'Jumbo'	0-6-0	539	2,067,787	567	1,364,674
812 class	0-6-0	812	1,574,248	661	1,002,170
300 class	0-6-0	318	783,147	323	580,616
40 class	0-6-0	41	1,626,225	46	1,178,926
Pickersgill	4-4-0	116	1,078,002	67	789,675
439 class	0-4-4T	442	1,892,805	426	814,218
782 class	0-6-0T	789	2,229,028	253	560,458

BIBLIOGRAPHY

The following have been consulted:—

(A) OFFICIAL RECORDS

(1) *Caledonian Railway*
Locomotive diagram books: 1903 and 1923
Locomotive Register: 1891-1917
St Rollox boiler-proving register: To 1897
St Rollox locomotive drawings
Specification for 812 class 0-6-0s to be built by Neilson, Reid & Co
Specification for 4-4-0 No 124 to be built by Dübs & Co
Locomotive department's file of extracts from minutes of board meetings and relative letters: 1872-1920
Locomotive department's expenditure: 1887 and 1890-1
Inventory of locomotive department's plant, machinery and furnishings, 1 October 1917
Rolling stock renewals: memoranda and extracts of minutes
Staff emoluments and duties: 1904-20
Individual staff histories: 1892-1932
Personal letter from J. F. McIntosh to D. H. Littlejohn
File of correspondence between J. F. McIntosh and Dr John Inglis
Minute books of directors and committees, selected
Mineral timetables, working timetables and appendices, selected

(2) *Highland Railway*
Minutes of directors: May 1914 to March 1917
Specification for River class 4-6-0s

(3) *LMS Railway*
Northern division boiler diagram book

(4) *Neilson & Co*
Diagram books

BIBLIOGRAPHY

(B) BOOKS

MacLeod, A. B. *The McIntosh Locomotives of the Caledonian Railway* (1948)
Poultney, E. C. *British Express Locomotive Development 1896-1948* (1952)
Ahrons, E. L. *Locomotive and Train Working in the Latter Part of the Nineteenth Century*: Volume Three (Reprinted 1952)
Nock, O. S. *The Railway Race to the North*
A Livery Register: The Caledonian Railway: Locomotives 1883-1923. The Historical Model Railway Society
Cox, E. S. *Chronicles of Steam* (1967)
Thomas, John. *The Story of 828*
Nock, O. S. *The Caledonian Dunalastairs* (1968)
Nock, O. S. *The Caledonian Railway*
Thomas, John. *The Callander & Oban Railway* (1966)
Thomas, John. *The Springburn Story* (1964)

(C) ARTICLES

*'Express Locomotive for the Caledonian Railway' (No 123), *Engineering*, 10 December 1886. See also *The Engineer*, 13, 20 & 27 August 1886
*'Passenger Locomotive for the Caledonian Railway' (No 124), *Engineering*, 30 July 1886
*'Bogie Passenger Engine, Caledonian Railway' (80 class), *The Engineer*, 13 April 1888
'Carriage Heating Apparatus, Caledonian Railway', *The Engineer*, 4 January 1889
*'Express Passenger Engine, Caledonian Railway' (13 class), *The Engineer*, 31 May 1895
Rous-Marten, C. 'The New Caledonian Express-Engine *Dunalastair*', *The Engineer*, 28 February 1896. See also *The Railway World*, May 1897
Rous-Marten, C. 'With the West Coast 'Flyer',' *The Engineer*, 21 August 1896
Drummond, D. 'An Investigation into the Use of Progressive High Pressures in Non-Compound Locomotive Engines'. *Minutes of the Proceedings of the Institution of Civil Engineers* 1897, Volume 127, Paper No 2497
Dunn, P. L. 'The St Rollox Locomotive and Carriage Works of the Caledonian Railway', *Minutes of the Proceedings of the Institution of Civil Engineers*. Volume 129, Paper No 2696
Rous-Marten, C. 'A New Express Engine on the Caledonian Railway' (766 class), *The Engineer*, 25 February, 1898
*'Caledonian Railway Passenger Locomotives' (including 900 class), *Engineering*, 18 May 1900
*'Six-Wheeled Coupled Passenger Locomotive: Caledonian Railway' (55 class), *Engineering*, 29 August 1902

BIBLIOGRAPHY

*'Bogie Tender: Caledonian Railway', *The Railway Engineer,* August 1902

'Six-Coupled Bogie Express Passenger Engine: Caledonian Railway' (49 class), *The Railway Engineer,* July 1903

'Six-Wheel Coupled Passenger Locomotive: Caledonian Railway' (49 class), *Engineering,* 18 September, 1903

Rous-Marten, C. 'A Caledonian Locomotive Giant' (49 class), *The Engineer,* 3 April 1903

Rous-Marten, C. 'The New Caledonian Giants at Work', *The Engineer,* 21 August, 1903

*'New 4-4-0 Type Locomotive Caledonian Railway' (140 class), *The Railway Engineer,* October 1905

*'Six-Coupled Passenger Express Locomotive: Caledonian Railway' (903 class), *Engineering,* 1 February 1907

McIntosh, J. F. 'British Express Locomotives', *Cassier's Magazine,* March 1910

Thompson, Thomas. 'Notes on Experiments on Locomotive Spark-Arrestors', *Transactions of the Institution of Engineers and Shipbuilders in Scotland,* 28 March 1910

*'4-6-0 Express Goods Engines, Caledonian Railway' (179 class), *The Railway Engineer,* April 1914

*'4-4-0 Express Passenger Engines, Caledonian Railway' (117 class), *The Railway Engineer,* June 1914

Kempt, Irvine Jnr. 'Locomotive Lubrication'. *Journal of the Institution of Locomotive Engineers,* March 1921

Poultney, E. C. 'Recent Locomotive Practice on the Caledonian Railway' (60, 113, 944 & 300 classes), *The Engineer,* 4 & 11 November 1921

'New Three-Cylinder 4-6-0 Type Locomotives, Caledonian Railway', *The Railway Engineer,* October 1921

Ward, L. 'Caledonian Railway: The Killin Branch Tanks', SLS *Journal,* March 1943

Macleod, A. B. 'The Lambie Locomotives of the Caledonian Railway', *The Model Railway News,* July, August & September 1944

MacLeod, A. B. 'Caledonian Railway Number-plates', *The Model Railway News,* September 1944

Newlands, D. 'The Caley 60 Class', SLS *Journal,* August 1946

McEwan, J. F. 'The Locomotives of the Caledonian Railway' (1883-95 period), *The Locomotive,* December, 1946, February, April, June, September, November & December 1947, March, May & June, 1948

Dunbar, A. G. 'Y125- And All That?', SLS *Journal,* August 1948

Keiller, C. M. 'Drummond Cylinders', SLS *Journal,* September 1953

Dunbar, A. G. 'The Caledonian 40 class 4-4-0s', *Trains Illustrated,* January 1954

BIBLIOGRAPHY

Dunbar, A. G. 'The Dunalastair I 4-4-0s of the Caledonian', *Trains Illustrated,* February 1959

Dunbar, A. G. 'The Cardeans of the Caledonian', *Trains Illustrated,* June 1959

Paterson, A. J. S. 'Exhibition Engines of 1886', *Railway World,* January 1960

Paterson, A. J. S. 'Dugald Drummond at St Rollox, 1882-85', *Railway World,* January & March 1961

Paterson, A. J. S. 'The McIntosh 0-6-0s and 2-6-0s of the Caledonian Railway', SLS *Journal,* November 1962

Dunbar, A. G. 'The McIntosh eight-coupled engines and Moguls of the Caledonian Railway', *Railway World,* January 1963

Dunbar, A. G. 'The Lambie 4-4-0 Tanks of the Caledonian Railway', *Railway World,* October 1963

Dunbar, A. G. 'The McIntosh 918 class 4-6-0s of the Caledonian', *Railway World,* May 1964

Paterson, A. J. S. 'Some Notes on the McIntosh 6ft 6in 4-6-0s of the Caledonian Railway', SLS *Journal,* September 1965 and February 1966

Dunbar, A. G. '50 Years ago at Balornock Shed', SLS *Journal,* July 1972

*denotes general arrangement drawing included

INDEX

(*Entries in italics refer to illustrations*)

Adams, William, LSWR locomotive superintendent, 30
Allen, C. J., 149
Allocation of locomotives, 22, 35, 36, 37, 38, 41, 42, 48, 55, 57, 61, 62, 74, 76, 83, 84, 89, 90, 92, 93, 94, 97, 98, 101, 107, 112, 121, 127, 128, 129, 130, 131, 140, 143, 145, 147, 149, 151, 152, 161, 164, 169, 173–4, 175, 176–7, 178, 179, 183, 186, 206–8
Arguzoid, 32
Armstrong Whitworth & Co, 179
Arran Express, 36
Ashpan design, 20, 45, 56, 100, 104, 111, 120, 127, 128, 147, 168, 185
Assistant locomotive superintendent, 15, 50, 68, 71, 155
Atlantic locomotives, 117, 205
Atlas Works, 163
Australasian Locomotive Engine Works Ltd, 49
Axlebox guides, 112, 119, 167, 176

Banking engines, 176
Barochan, 128
Barr, J. G., 117, 155
Balanced side-valves, 110, 118, 173, *173*, 185
Beattock: coal-handling facilities, 175

Blastpipe design, 20, 30, 35, 38, 45, 55, *58*, 59, 60, 62, 72, 80, 89, 96, 100, 104, 110, 111, 120, 129, 130, *131*, 132, 135, 164
Board minutes, 14, 29–30, 49, 117, 155, 156
Bogie design, 22, 55, 72, 75, 93, 102, 112, 161, 163, 168, 176, 182
Boiler design, 19, 26, 30, 38, 45, 54, 55, 59, 60, 75, 80, 96, 99, 100, 102, 104, 110, 111, 113, 119, 121, 126, 131, 132, 151, 158, 163, 168, 174, 182, 185, 196–8
Boiler repairs, 64
Boiler-proving register, 42, 45–6
Bolten, J. C., CR Chairman, 48
Bowen-Cooke, C. J., LNWR CME, 125
Bradshaw, 71
Brake arrangements, 19, 22, 24, 27, 41, 42–3, 58, 59, 60–2, 63, 71, 75, 81, 83, 99, 128, 151, 161, 164
Breadalbane, 83, 176
Brittain, George, 13, 14, 49
Bryce-Douglas valve gear, 32–3
Buffers, 27, 41, 42, 61, 164
Bunten, J. C., CR Chairman, 74
By-pass valves, 163, 165, 167, 180

Cabs, 16, 26, 27, 40, 42, 60, 75, 80, 83, 90, 93, 97, 104, 112,

215

INDEX

113, 127, 128, 131, 146, 161, 169, 174, 183, 202–3
Cab fittings, 20, 23, 24, *25*, 26, 27, *31*, 32, 38, 40–1, 45, 61, 62, 63, 64, 80, 90, 93, 104, 121, 127, 128, 129, *139*, 140, 148, 164, 169, 174, *182*
Caledonian mechanical lubricator, 164, 169, 172, 182, 185
Campbell, Archibald, NBR chief draughtsman, 157
Capital value of locomotives, 115–16, 117, 136–7
Carbonisation of oil in cylinders, 153, 177
Carbrook, 48
Cardean, 121, 125, 148
Carriage warming apparatus, 35, 38, 45, 60, 67, 128, 150, 174, 177
Cassier's Magazine 153–4
Chief draughtsman, 15, 71, 78, 179
Chimneys, 20, 21, 30, 35, 75, 83, 158, 164
Coal consumption, 43, 48, 56–7, 82, 127, 140, 147–8, 178–9
Coal prices, 156
Compound locomotives, 43, 205
Compressed air sanding, 30, 38, 45, 113
Condensing apparatus, 62, 63, 72, 177
Connecting rods, 18, 35, 83, 100, 102, 111, 118, 119, 150, 167, 180, 200–1
Conner, Benjamin, 49
Consolidated Carriage Company, 67
Consolidated superheater, 143, 146, 149
Cost of engines, 21, 22, 23, 26, 30, 38, 41, 42, 43, 44, 50, 54, 55, 58, 59, 60, 61, 63, 71, 72, 73, 74, 79, 84, 85, 89, 92, 93, 94, 95, 99, 101, 111, 112, 115, 117, 126, 129, 130, 138, 143, 144, 145, 146, 150, 151, 152, 157, 158, 162, 163, 165, 172, 175, 176

Coupling rods, 96, 111, 119, 126, 167
Cox, Edward, Deputy Chairman, 121
Crank-axles, 19, 21, 22, 23, 38, 41, 50, 58, 80, 83, 92, 102, 111, 118, 126, 130, 140, 145, 146, 150, 163
Crewe: method of flanging firebox, 104
Crosshead pumps, 61, 64
Currie, Driver James, 90, 121, 148
Cylinders, 18, *19*, 24, 26, 30, 35, 38, *44*, 44–5, 48, 50, *53*, 55, 59, 61, 74, *79*, 79, 83, 92, 95, 101, 110–11, 118, 132, 139, 143, 144, 146, 148, 149, 150, 158, 163, 165, 172, 174, 179, 180, 185, 199–201

Davies & Metcalfe exhaust steam injector, 178
Derived valve gear, *181*, 184
Detroit sight-feed lubricator, 164, 178
Dome, 19, 54, 55, 59, 75, 80, 83, 158
Dragbox, 121, 163, 169, 172, 183
Drop-grate, 157, 158
Drummond, Dugald, 14, 27–8, 29, 45, 48, 49
Drummond, Peter, 15, 71
Drummond injectors, 19, 24, 30, 40, 42, 45, 61, 64, 72, 73
Dübs & Co, 29, 32, 35, 85
Dunalastair, 74
Dunalastair 2nd, 83
Dunn, Driver Andrew, 81, 90

Eccentric-rods, 18, 22, 45, 55, 63, 72, 75, 79, 83, 95, 100, 101, 102, 118, 140, 145, 146, 149, *166*, 167, 200–1
Eccentric-sheaves, 101, 148, *166*, 167
Edinburgh International Exhibition: 1886; 29, 35
Edinburgh International Exhibition: 1890; 48

216

INDEX

Engineering, 91
Exhaust steam injector, 178
Expansion links, 18, 150, 167

Fairfield Shipbuilding & Engineering Co, Govan, 32
Finance Committee, 154
Firebox design, 19, 26, 42, 54, 55, 59, 75, 89, 100, *102*, 104, 111, 113, 119–20, *120*, 127, 129, 132, 147, 151, 158, 163, 168, 174, 182, 185, 196–8
Flanging blocks, 182
Footsteps, 50, 55, 56, 65, 71, 74, 76, 129
Frames, 22, 26, 30, 35, 38, 42, 55, 61, 75, 79, 80, 96, 99, 102, 119, 131, 140, 161, 163, 167, 172, 174, 185, 200–1
Friedmann mechanical lubricator, 140
Furness lubricator, 56, 62, 65, 76, 84, 104, 111–12

General Manager, 14
General Manager's memo on renewals, 137
Gibson, Driver David, 148, 169
Glasgow Central Railway, 61, 72
Glasgow International Exhibition: 1888; 36
Glasgow & Paisley Joint Line Committee, 41
Goodfellow, Joseph, 15, 71
Governor for Westinghouse pump, 80, 93
Grampian Corridor Express, 128, 140, 146
Grassie, Driver James, 183
Gravity sanding, 35, 41, 50, 55, 61, 63, 75, 83, 96, 100, 119, 130, 161
Gresham & Craven steam sanding, 61, 80
Gresham & Craven combination injectors, 72, 75, 83, 113, 164, 169, 185
Gresham & Craven vacuum ejector, 161, 164, 169

Hammerblow, 161
Handrails, 20, 50, 55, 56, 59, 60, 65, 76
Handrail-plate (tender), 59, 60
Hawthorn Leslie & Co, 156, 157
High-capacity bogie wagons, 95
Highland Railway, 156, 157–8
Holden system of oil-burning, 150
Holt, F., Derby Works Manager, 30
Hood, 75, *103, 109*, 129, *133, 134*, 135
Hyde Park Works, 179

Inglis, Dr John: North British Railway, 115–16
Injectors, 19, 24, 30, 40, 42, 45, 61, 64, 72, 73, 75, 83, 104, 113, 158, 164, 169, 178, 185
Institution of Civil Engineers, 43
Institution of Locomotive Engineers, 177
Irvine Kempt, Junior, 117, 155, 177

Jeffrey, James: Painter, St Rollox, 107
Journals (locomotive), 32, 80, 96, 102, 112, 119, 126, 145, 146, 161, 163, 167, 181, 200–1
Journals (tender), 39, 40, 65–6, 164
Jubilee, 78
'Jubilee Pugs', 41

King, Sir James, 68, 128

Lambie, John, 15, 50
Lanarkshire & Dumbartonshire Railway, 61
Livery, 21, 32, 35, 36, 42, 62, 74, 78, 83, 85–6, 107, 121, 128, 146, 179
Locomotive classes:
  Conner: 15–16
    Rebuilt 7ft 2in 2–4–0s, 101, 114
  'Rebuilds', 16–17
  Brittain: 17, 29

INDEX

Oban bogies, 99
670 class 0-4-2s 99
Drummond:
'Jumbos' 18-22, 36, 39, 40, 58, *59*, 60, 65, 66, 71-2, 97, 99, 156, 171, 177, 189-91
66 class 4-4-0s 22-4, 39, 44-8, 50-4, 56-7, 65, 66, 98-9, 132, 156, 187
171 class 0-4-4Ts 24, *25*, 26, 43, 60, 131, 132, 135, 192
80 class 4-4-0, 37-8, 39, 40, 54-5, 75, 78, 99, 132, 176, 177, 187
385 class 0-6-0ST 40-2, 194
272 class 0-6-0ST 42, 193-4
262 class 0-4-2ST 26-7, 192
264 class 0-4-0ST 26-7, 41-3, 72, 94, 130, 132, 192
123 class 4-2-2 29-30, 32, 35-7, 39, 43, 187
124 class 4-4-0 32-3, 36-7, 39
Lambie:
1 class 4-4-0T 61, 177, 192
211 class 0-6-0ST 60, 194
19 class 0-4-4T 63, 132, 177, 193
13 class 4-4-0 55-6, 57, 65, 67, 132, 187-8

McIntosh:
29 class 0-6-0T 72, 132, 177, 194
92 class 0-4-4T 73, 132, 177, 193
879 class 0-4-4T 73, 132, 193
721 class 4-4-0 74-7, 78, 82, 131, 132, 150, 176-7, 188
766 4-4-0 78-83, 149, 176-7, 188
812 class 0-6-0 83-9, *84*, 97, 132, 150, 191
900 class 4-4-0 89-92, 149, 177, 178, 188
782 class 0-6-0T 92, 129, 132, 135, 151-2, 175-6, 194-5

104 class 0-4-4T 92-3 *131*, 132, 193
439 class 0-4-4T 93-4, 108, 130, 132, *135*, 152, 176, 193
600 class 0-8-0 95-8, 192
55 class 4-6-0 99-101, 135, 156, 191
49 class 4-6-0 101-11, *102*, *103*, *109*, 148, 176, 191
492 class 0-8-0T 111-12, 135, 195
140 class 4-4-0 112-15, 130, 132, 133-4, *133*, *134*, 137-8, 149, 150, 177, 178, 188
Proposed 'Atlantics' 117, 205
903 class 4-6-0 *102*, 117-26, *118*, *120*, 148, 177, 191
918 class 4-6-0 126-7, 191
908 class 4-6-0 127-8, 132, 146, 149, 177, 191
652 class 0-6-0 130-1, 132, 191
139 class 4-4-0 137-40, 188
117 and 40 class 4-4-0s *139*, 143-4, 188
30 class 0-6-0 144-5, 191
34 class 2-6-0 145-6, 191
179 class 4-6-0 146-8, 192
498 class 0-6-0T 150-1, 175, 195
Proposed 4-6-0s 154, 165
Proposed 'Pacifics' 154, 205
Pickersgill:
113 and 72 class 4-4-0s 162-5, 179, 188-9
60 class 4-6-0 165-70, *168*, 176, 192
Proposed 2-6-0s 172, 186, 205
300 class 0-6-0 172-4, *173*, 191
944 class 4-6-2T 174-5, 195
956 class 4-6-0 180-4, *180*, *181*, *182*, 192
191 class 4-6-0 184-6, 192
Others:
ex-HR 'River' class 156-62, 192
Neilson & Co 0-4-0STs 26
LNWR 'Precursors' 82

INDEX

LNWR 0-8-0 179
Midland 'singles' 82
NBR classes, 18, 24
NER Fletcher 0-6-0s 156
GCR 0-6-0s 156
GCR 2-8-0s 179
SR *King Arthur* 183-4
Gresley Pacific 184
Locomotive expenditure, 27-8, 115-16, 136-7
Locomotive register, 22, 23, 38, 39, 171
Locomotive & Stores Committee, 13, 14, 26, 68, 126, 155, 171, 175
Locomotive Superintendent, 13, 14, 50, 68, 155
London & North Western Railway, 125
Lubrication, 20, 40, 56, 62, 80, 83, 84, 100, 104, 113, 121, 126, 129, 140, 143, 145, 146, 148, 149, 158, 164, 169, 172, 174, 177-8, 179, 182, 185

MacDonald, Thomas, 71, 107, 117
McIntosh, J. F., 50, 68, 107, 116, 132, 153-4, 155
McIntosh patent water-level gauge protector, 75
Mackie, Driver George, 169, 178
Mechanical lubricator, 140, 143, 145, 146, 148, 149, 158, 164, 169, 172, 174, 177, 178, 182, 185
Menno grease lubricators, 179
Metallic lubricator, 40
Metallic packing, 149
Mitchell, Driver 'Cuddy', 114
Monkswell, Lord, 77, 91
Moodie, W. H., 179
Motion, 38, 119, 150, 163, 167, 180, 184
Motion-plate, 18, 22, 30, 35, 38, 45, 55, 61, 63, 72, 75, 79, 96, 101

Neilson & Co, 21, 23, 26, 29, 30, 32, 41

Neilson, Reid & Co, 85
'New Jubilees', 61
Nock, O. S., 57, 77, 125, 164-5, 170
North British Locomotive Company, 156, 163, 175, 179, 182, 185
North British Railway, 14, 28, 32, 39, 115-16, 157
Numberplates, 73, 85, 98, 161, 164, 175, 179

Oil burning, 150
Oil painting of No 50, 107

Patrick, William, Assistant General Manager, 50
Patents, 132, 149
Pearson Pattinson, J., 36, 53-4
Performance, 35-7, 43, 46-8, 53-4, 56-7, 77, 81-2, 91, 107-8, 110, 113-15, 122, 125, 128, 140, 147, 149, 153, 164-5, 169-70, 183-4
Pet cock, 62
Petticoat pipe, 89, 129, *103, 109, 133*
Pickersgill, William, 155
Piloting trains, 24, 91, 101, 144, 153
Piston valves, 137, 140, 144, 146, 148, 149, 158, 163, 167, 172, 181
Pony truck, 145, 174
Poultney, E. C., 47-8, 82, 170
Preston, W. R., 132
Prince of Wales feathers, 36
Pyrometer, 140, 148, 169

Queen's Park Works, 185
Quintinshill disaster, 162

Race to Aberdeen, 56-7
Race to Edinburgh, 35
Railway Executive Committee, 171
Railway Operating Division (ROD), 179
Ramsbottom double-beat regulator, 55
Ramsbottom safety-valves, 19
Ranochan, Driver J., 76, 90, 107

INDEX

'Rebuild' boiler, 40, 60, 61, 72, 92, 99
Rebuilding, 16–17, 98–9, 131–2, 148–9, 178, 184
Regulator valve, 19, 55, 158
Renewal of locomotive stock, 115–16, 136–7
Renshaw, Sir Charles Bine, M.P., 128
Reversing quadrant, 20, 23, 24, 27, 30, 38, 55, 75, 80, 83, 90, 96, 104, 113, 140
Richardson balanced slide-valves, 110, 118
Robert Stephenson & Co, 156
Robinson draught-retarder, 143
Robinson Intensifore lubricator, 177
Rocking lever, 95, 101, 140, 144, 146, 149, 166
Ross pop safety valves, 158, 164, 177, 179, 182, 185
Rous-Marten, C., 57, 76, 77, 81, 82, 91, 107–8, 110, 113
Royal Train duty, 36, 39

Safety-valves, 19, 55, 59, 96, 100, 104, 111, 113, 121, 130, 176, 177, 179, 182, 185
Salaries, 14, 28, 49, 50, 68, 71, 117, 154, 155
Sanding, 24, 30, 35, 38, 41, 45, 50, 55, 61, 63, 80, 93, 100, 104, 113, 119, 130, 161, 164, 172, 174, 176, 177, 183, 185
Schmidt superheater, 137, 138, 144, 148
Scottish miners' strike: 1894; 71
Scottish North Eastern Railway, 68
Scottish Rail strike: 1890–1; 50
Sharp, Stewart & Co, 85
Sight-feed lubricator, 56, 62, 80, 83, 93, 129, 164, 177, 178
Sighthill cemetery, 71
Slidebars, 18, 22, 24, 30, 35, 38, 45, 55, 61, 63, 71, 75, 79, 167
Slide-valves, 18, 45, 79, 110, 173, *173*, 185

Smellie, Hugh, 50
Smith, F. G., 157
Smith feed-water heater, 158
Smokebox, 20, 56, 63, 80, 83, *103*, *109*, 110, *133*, *134*, 139, 144, 169, 172
Snifting valve, 140, 169, 179, 182
Snowball, Edward, 29
Spark arrestor, 132–3, *134*, 138–9
Spencer, John & Sons Ltd, 125
Springs, 19, 22, 24, 63, 96, 100, 102, 111, 112, 131, 140, 151, 158, 168, 172, 174, 179, 182–3, 184, 185
Stavert, Driver William, 114, 121
Steam heating, 35, 38, 45, 60, 67, 128, 150, 174, 177
Steam reverser, 90 100, 104, 121, 128, 140, 161, 169, 174, 181, 185
Steam sanding, 61, 80, 104, 113, 164, 172, 174, 176, 177, 183, 185
Steam sanding and blower valve, 164, 169, 183
Steam temperature, 140
Steel fireboxes, 42, 172
Stone & Co, Deptford, 132
Stone's mechanical lubricator, 143
St Rollox, 13, 136
Superheater damper, 138, 143, 144, 146, 148, 149, 164
Superheater elements, 138, 143, 144, 146, 148, 149, 158, 163, 168, 174, 182

Tablet exchanging apparatus, 100
Tenders, 20–1, 22, 23, 32, 35, 36, 38, 39–40, 45, 53, 55, 56, 58, 59, 60, 65, 71, 75, 81, 84, *85*, 90, 97, 100, 104, 113, 137, 143, 150, 164, 169, 172, 173, 177, 183, 186, 204
Tender cabs, 177
Tender water-level gauge, 72, 97
The Arran Express, 36
The Caledonian Dunalastairs: O.S. Nock, 77, 164
The 'Corridor', 56, 76, 77, 81, 82, 90, 91, 107, 108, 121, 149, 169, 183

220

INDEX

The Engineer, 38, 57, 107, 110, 113, 170
The Railway Magazine, 77, 149, 176
The Tinto Express, 177
The 'Tourist', 76-7, 82
Thompson, Sir James, CR General Manager, 50, 116
Todd, Driver William, 107
Top-feed apparatus, 150
Traffic and permanent way committee, 14, 157, 162
Train heating, 35, 38, 45, 60, 67, 128, 150, 174, 177

Urie, Robert Wallace, 15, 71, 78
Urie, William Mongomerie, 15, 71, 78, 155

Vacuum brake, 59, 66, *66*, 84, 90, 104, 130, 131, 145, 146, 152, 161, 164, 169, 174, 177, 183
Valve chests, 18, *19*, 35, *44*, 44-5, 48, 50, *53*, 55, 61, 74, *79*, 83, 95, 101, 139, 144, 146, 150, 158, 165, 180, 185, 199
Valve gear, 18, 22, 30, 32-3, 38, 45, 55, 63, 72, 75, 79, 80, 83, 95, 100, 101, 118, 140, 144, 146, 148, 149, 150, 158, *166*, 167, 174, *181*, 184
Valve setting, 96, 102, 138, 144, 150, 165, 173, 181, 184, 185, 199
Victoria, 78
Vortex blastpipe, 30, 38, 45, 55-6, 60, 72, 75, 83, 92

Wakefield mechanical lubricator, 143, 146, 148, 158, 174
Walschaerts valve gear, 158, 181, 185
War service, 171
Water consumption, 56-7, 82, 140, 144, 178
Water-level gauge, 72, 97
Watt, driver, 114
Weir feedwater heater and pump, 150
Weir, Thomas, 78, 179
'Wemyss Bay tanks', 175
Westinghouse brake system, 22, 27, 32, 35, 38, 60, 63, 66, *66*, 71, 75, 80, 93, 112, 113, 177, 183
Westinghouse feed pump, 72, 73
Wheatley, George Thomas, 15
Wheels, 19, 26, 40, 60, 72, 75, 96, 111, *118*, 119, 126, 145
Whistles, 30, 35, 38, 56, 60, 149
Whitelaw, William, 157
Works manager, 15, 71, 78
Worthington, Edgar, 43

York, Colonel, 125-6
Yorkshire Engine Co, 156